Belize Audubon Society:
40 Years of Working for a Balance between People and the Environment

(*frontspiece photo*) The entire BAS Board and Staff met at Chaa Creek in western Belize for a retreat to develop the Strategic Plan for 2008-2013.

(*standing left to right*) John Hunt (*invited facilitator*), Israel Manzanero Sr., Annette Reneau, Nicole Auil, Marco Leal, Mike Heusner, Federico Tush, Kenneth Gale, Carlos Matus, Rennie Jones, Lascelle Tillet, Earl Green, Ivis Chan, Arvin Coc, Kent Garcia, John Arana, Carol Williams, Carlton Young, Israel Manzanero Jr, Felicia Leal, Anna Hoare, Roberto Pott, Nicacio Coc, Khadia Mosiah, Dominque Lizama, Daniel Chan, Mark Macfield, Orlando Jimenez, John Searle, Uriah Tillet, Nikisha Ack, Carlos Santos, Linda Searle, Shaun Finetty (*invited facilitator*)

(*below left to right*) Kristin Hunt (*invited facilitator*), Juan Chun, Elisa Ariola, Albart Liconia, Marcelo Pau, Mario Bolon, Anicacio Garcia, Derick Hendy, Brijilio Bolon, Bonifacio Cal, Harvey Rosado, Olivia Carballo, Jairo Juchim, Scott Schimdt, Norman Mora, Transito Jimenez, Ignacio Pop.

Belize Audubon Society:
40 Years of Working for a Balance between People and the Environment

Judy Lumb and Lydia Waight

Photographs by Dirk Francisco

Producciones de la Hamaca

Caye Caulker, BELIZE

2011

BELIZE AUDUBON SOCIETY

Celebrating 40 Years

Published for the Belize Audubon Society
by *Producciones de la Hamaca*
Caye Caulker, BELIZE
<producciones-hamaca.com>

ISBN: 978-976-8142-29-0

Cover designed by Dirk Francisco
Unless otherwise indicated, all photos are by Dirk Francisco

BAS Vision

The Belize Audubon Society (BAS) will be a strategic leader in building a nation in which people in Belize live in harmony with and benefit from the environment.

BAS Mission

The Belize Audubon Society is a non-governmental membership organization dedicated to the sustainable management of our natural resources through leadership and strategic partnerships with stakeholders in order to achieve, create and maintain a balance between people and the environment.

Producciones de la Hamaca is dedicated to:

- Celebration and documentation of Belize's rich, diverse cultural heritage,
- Protection and sustainable use of Belize's remarkable natural resources,
- Inspired, creative expression of Belize's spiritual depth.

Contents

Forewords from BAS Presidents viii
Acknowledgements x

Chapter One—Formation and Growth of the Belize Audubon Society 1

An Introduction to the Founders 2
Field Trips 5
The Early Years 7
First Annual General Meeting 7
Becoming Independent 7
First BAS Logo 8
Updated BAS Logo 8
The First Fifteen Years 9
A Time of Change 9
BAS Administrative Staff 11
BAS Membership 15
BAS Organization 17
Annual General Meetings 18
Cooperation with the Government 25
BAS Workshops and Retreats 26
Strategic Plans 29
Financial Trials 33
BAS Office 35
25th Anniversary Celebration 39
30th Anniversary Celebration 40
40th Anniversary Celebration 41
Summary—Formation and Growth of the Belize Audubon Society 49

Chapter Two—The Advocacy Programme at the Belize Audubon Society 51

National Parks and Wildlife Protection Legislation 51
Wildlife Protection 52
Protecting Marine Resources 55
Pesticides 57
National Conservation Strategy 57
National Protected Areas Policy and System Plan Project 57
Game Ranch Project 57
Coca Cola 58
Logging in Toledo District 59
National Environmental Appraisal Committee 61
Other Committee Membership 61
Lamanai Room Declaration 62
Lamanai Room Declaration II 63
HIVOS Grant 65
An Environmental Agenda for the 21st Century 65
Environmental Agenda 2002 and Beyond 66
An Environmental Agenda for Belize 2008-2013 66
Chalillo Hydroelectric Dam 67
Land Policy Reform 68
Tilapia in Crooked Tree Lagoon 69
Fish Kill in Crooked Tree Lagoon 70
Threats to Crooked Tree Wildlife Sanctuary 71
Cattle Farming Best Practices for Crooked Tree Wetland Conservation 73
BAS on Mega-Developments 75
Protecting Belize's Water Resources 76
Waste Management 76
Earth Summit 76
Oil Refinery Planned for Monkey River 77
Oil Exploration in Sarstoon-Temash National Park 77
Offshore Drilling for Oil 78
BAS Media Campaigns 79
40 years of BAS Advocacy 80

Chapter Three—Environmental Education 81

Goals and Objectives of Environmental Education 81
Youth Education Committee 81
Coastal Zone Management Project 83
RARE Programme 84
Environmental Conservation Carnivals 84
Training Programmes 85
Environmental Education Programme 86
Training for BAS Staff 87
Nature Schools 89
Training for Teachers 90
Exhibitions 93
Earth Day Celebrations 95
World Wetlands Day 99
CTWS Bird Festival 102
Junior Bird Club 103
Environmental Youth Camps 104
Educational Materials 107
BAS Books 109
Brochures 112
Colouring/Comic Books 114
Posters 115
Slide Shows and Videos 117
Filming at BAS-Managed Parks 118
BAS on the Web 119
Summary of BAS Environmental Education 122

Chapter Four—Protected Areas 123

Management of Protected Areas 123
Half Moon Caye Natural Monument 124
Conch Bust 130
Blue Hole Natural Monument 131
Promoting Conservation through Sustainable Tourism Development 132
Crown Reserve Bird Sanctuaries 134
Columbia Forest Wildlife Refuge 134
Bladen Branch Nature Reserve 134
Guanacaste National Park 135
St. Herman's Blue Hole National Park 140
Tapir Mountain Nature Reserve 146
Actun Tunichil Muknal Natural Monument 148
Crooked Tree Wildlife Sanctuary 150
Cockscomb Basin Wildlife Sanctuary 158
Victoria Peak Natural Monument 169
Summary of BAS Management of Protected Areas 170

Chapter Five—Research for Biodiversity and Conservation 171

BAS Research for Biodiversity and Conservation 171
Christmas Bird Counts 171
Important Bird Areas and Christmas Bird Count Areas 173
Other Bird Counts and Festivals 175
Jabiru Studies 178
Bird Observations in the *BAS Newsletter* 179
BAS Biodiversity Monitoring 180
Research in Crooked Tree 182
Research in Cockscomb 184
Cockscomb Basin Mammal Survey 185
Jaguar Research 186
Black Howler Monkey Translocation 188
Victoria Peak Amphibian and Biodiversity Expedition 190
Collaborative Assessment for Amphibian Conservation 191
Plant Inventory in CBWS 192
Xaté Project 194
Marine Forum 195

Limits of Acceptable Change for the Blue Hole and Half Moon Caye Natural Monuments 196
Marine Research 197
Spawning Aggregation 198
BAS Staff Research Results 198
Monitoring at Half Moon Caye and Blue Hole Natural Monuments 199
Sea Grass Monitoring 2006–2010 199
Lobster and Conch Monitoring at Lighthouse Reef 200
Sea Turtle Monitoring 201
Lionfish Eradication 201
Lighthouse Reef Marine Research 202
Summary of Research for Biodiversity and Conservation 204

Chapter Six—Awards 205

James A. Waight Conservation Award 205
World Wildlife Fund Award 222
Warden of the Year 222
Volunteer of the Year 223
Other BAS Awards 224
Belize National Awards 225
Goldman Environmental Prize 225
Knight of the Golden Ark 225
Summary of Awards 226

Chapter Seven—In Appreciation of Our Partners 227

Belize Center for Environmental Studies—Toledo Institute for Development and Environment 227
Green Reef-Belize 228
Community Baboon Sanctuary 228
Shipstern Nature Reserve 228
Belize Zoo 228
Programme for Belize 228
Belize Alliance of Conservation Non-government Organizations 229
Association of National 229
Development Agencies 229
Association of Protected Areas Management Organizations 229
Other Belizean Organizations 230
International Organizations 232
Volunteers 238
Stakeholder Communities 243
BAS as Co-manager 252
Funding Agencies 253
Major Supporters 253
Fund-raising Campaigns 260
Past and Present: Officers, Directors and Trustees 261
A Gallery of BAS Presidents 261
Thanks to all our BAS Board Members 262
Thanks to our BAS Staff 264
Summary and Conclusions 266

Index 267

FOREWORDS FROM BAS PRESIDENTS

When I returned to Belize from university in 1976 and started work at the Fisheries Department, I wanted to become more involved in marine conservation and also become more knowledgeable about Belize's natural resources and environment generally. BAS was very active at the time in bird-watching and environmental protection, so my husband and I joined as members.

When I was president, the Society was still working mostly on a voluntary basis. However, during my short one-year term we were able to strengthen and expand the paid Staff. I also think I was able to increase BAS' role in marine conservation, as I had previously chaired the Reef Preservation Committee for a few years.

BAS has made tremendous strides in protected area management, environmental education and public awareness. However, I think it's most important contribution now is, and will be, in the area of advocacy. The Society needs to keep focused on its environmental agenda and be a strong voice on how development should proceed to be truly environmentally sustainable.

Unfortunately at this time I am not very optimistic about the future of the environment in Belize. I feel Belize has lost its way to a large extent and has lost credibility internationally as an environmentally conscious nation.

We need to make a concerted effort to regain lost ground and regain our former recognition. This is made more difficult, of course, under the present global financial crisis. Despite this, though, Belize can make the right decisions and get back on track without necessarily losing financially.

—Janet Gibson
BAS President 1990–1991

I returned to Belize in 1981 after being away for five years. At the time BAS coordinated monthly birding field trips, produced a monthly newsletter, and hosted regular presentations at the Bliss Institute on Southern Foreshore. My parents had a family membership in BAS. I decided to take up bird watching as my hobby and started to attend the monthly field trips.

During my term of service on the Board and as President, BAS was expanding and consolidating its role in managing protected areas. Voluntary efforts were not enough for this task; more financial, physical and human resources were needed for this responsibility. I helped to recruit the first professional employees. The office space on Regent Street had become too small for the growing organization. We advocated for GOB support and succeeded in acquiring the BAS property at the Fort. We developed the first mission statement that recognized and stated the importance of the balance between people and the environment. The following are BAS contributions:

- BAS advocated for protected areas and wildlife protection that resulted in the enactment of the Protected Areas System and Wildlife Protection Acts.
- BAS initiated environmental education and awareness in the schools.
- BAS facilitated many research efforts and disseminated the resulting knowledge.
- BAS introduced many Belizeans, young and old, to conservation issues and encouraged appreciation for our marine and terrestrial resources.
- BAS helped the government make decisions on many environmental, conservation, and natural resource issues.

—Philip Balderamos
BAS President 1991–1992

Forewords from BAS Presidents (cont'd)

I became involved in BAS because of the voluntary work that the old original BAS folks were doing with only a love for nature and birding. In 1973 I came to Belize City to St. Johns College Sixth Form and became very good friends with Joey Waight and by extension with Mr. and Mrs. Waight.

BAS' important contributions during my term as President were growth and advocacy. We stood for what we believed, searching for a balance between people and nature. We stopped several projects that were not viable for Belize. The Lamanai Room Declaration was our climax and very important for our development in advocacy. Legalizing our status as national park managers was important for our development.

The general public always wanted to know what BAS thought about development and environmental issues. We advised the government of the day on the environment, national development, private and public projects, parks creation, legislation, PACT creation, Ministry of the Environment creation, and others too many to list. We served on many Boards because we were looked at as an important arm for a balanced national development.

The future will be challenging as more development occurs in Belize. In order to create a balance, the advocacy arm of BAS will have to strengthen and become more proactive. Too much anthropogenic sources of pollution are occurring today and it will increase as development and population increases in our country. BAS, along with the NGO community, will have to spend more time in developing strategies for education and advocacy in these areas.

—Jose Garcia
BAS President 1995–2000

Acknowledgements

This documentation of BAS' remarkable history would not have been possible without the untiring work of Lydia Waight. Her meticulous maintenance of records and sharp, insightful memory of the events of the first thirty years were the basis for the first history of BAS, *Belize Audubon Society: The First Thirty Years.*

Ms. Lydia was the Society's first Secretary and continued in that position for 27 years when she became Honorary Secretary for Life. She was the one who wrote all the letters for the early lobbying efforts of the BAS and kept the minutes of Board Meetings. It was my great privilege to work with her in the production of that first history of BAS. This updated history includes much of the original material in *Belize Audubon Society: The First Thirty Years*, albeit summarized.

It was only possible to make this updated history of the first 40 years of BAS into a full-colour coffee table book because of the photographic talent of Dirk Francisco. Dirk began his career at BAS as a volunteer in 2000. In 2002 the position of BAS Publicity and Public Participation Coordinator was created and he filled that important position. For ten years he has been documenting BAS activities and has amassed a huge portfolio of beautiful photographs, many of which are in this book.

Many people have contributed to this book by contributing information, reviewing the manuscript and a draft copy of the book, specifically, Amanda Burgos Acosta, David Craig, Meg Craig, Dirk Francisco, Earl Green, Anna D. Hoare, Losita Lee, Dominique Lizama, Linda Searle, and Lydia Waight. Thanks to all!

—Judy Lumb
June 26, 2011

Abbreviations

APAMO	Association of Protected Areas Management Organizations
ATMNM	Actun Tunichil Muknal Natural Monument
AVINA	A Swiss charity founded by Stephan Schmidheiny
BACONGO	Belize Alliance of Conservation Non-government Organizations
BAHA	Belize Agricultural Health Authority
BCES	Belize Center for Environmental Studies
BCVI	Belize Council for the Visually Impaired
BEST	Belize Enterprise for Sustainable Technology
BETEX	Belize Tourism Expo
BTIA	Belize Tourism Industry Association
BYCC	Belize Youth Conservation Corps
CATIE	*Centro Agronomico Tropical de Investigacion y Enseñanza*
CBWS	Cockscomb Basin Wildlife Sanctuary
CITES	Convention on International Trade of Endangered Species of Flora and Fauna
COMPACT	Community Management of Protected Areas for Conservation
CTWS	Crooked Tree Wildlife Sanctuary
DEMA	Diving Equipment and Marketing Association
DOE	Deparment of Environment
EIA	Environmental Impact Assessment
GOB	Government of Belize
HIVOS	Humanitarian Institute for Cooperation with Developing Countries
HMCNM	Half Moon Caye Natural Monument
ICTF	International Tropical Conservation Foundation
IoA	Institute of Archeology
IUCN	World Conservation Union (originally International Union for Conservation of Nature)
NARMAP	Natural Resources Management Project
NBC	National Broadcasting Corporation (U.S.)
NEAC	National Environmental Appraisal Committee
NGO	non-Governmental Organization
NICH	National Institute of Culture and History
NRDC	Natural Resources Defence Council
PACT	Protected Areas Conservation Trust
RARE	Rare Animal Relief Effort
SEED	Students Encouraging Environmental Development
SHBHNP	St. Herman's Blue Hole National Park (formerly Blue Hole National Park)
SPEAR	Society for the Promotion of Education and Research
SUN	Students Understanding Nature
SWOT	Strengths, Weaknesses, Opportunities, and Threats
TIDE	Toledo Institute for Development and Environment
TMNR	Tapir Mountain Nature Reserve (formerly Society Hall Nature Reserve)
TNC	The Nature Conservancy
UNESCO	United Nations Educational, Scientific and Cultural Organization
USAID	U.S. Agency for International Development
WCS	Wildlife Conservation Society
WWF	World Wildlife Fund
YEAG	Youth Environmental Action Group
YMCA	Young Men's Christian Association
YWCA	Young Women's Christian Association

Forty Years of the Belize Audubon Society

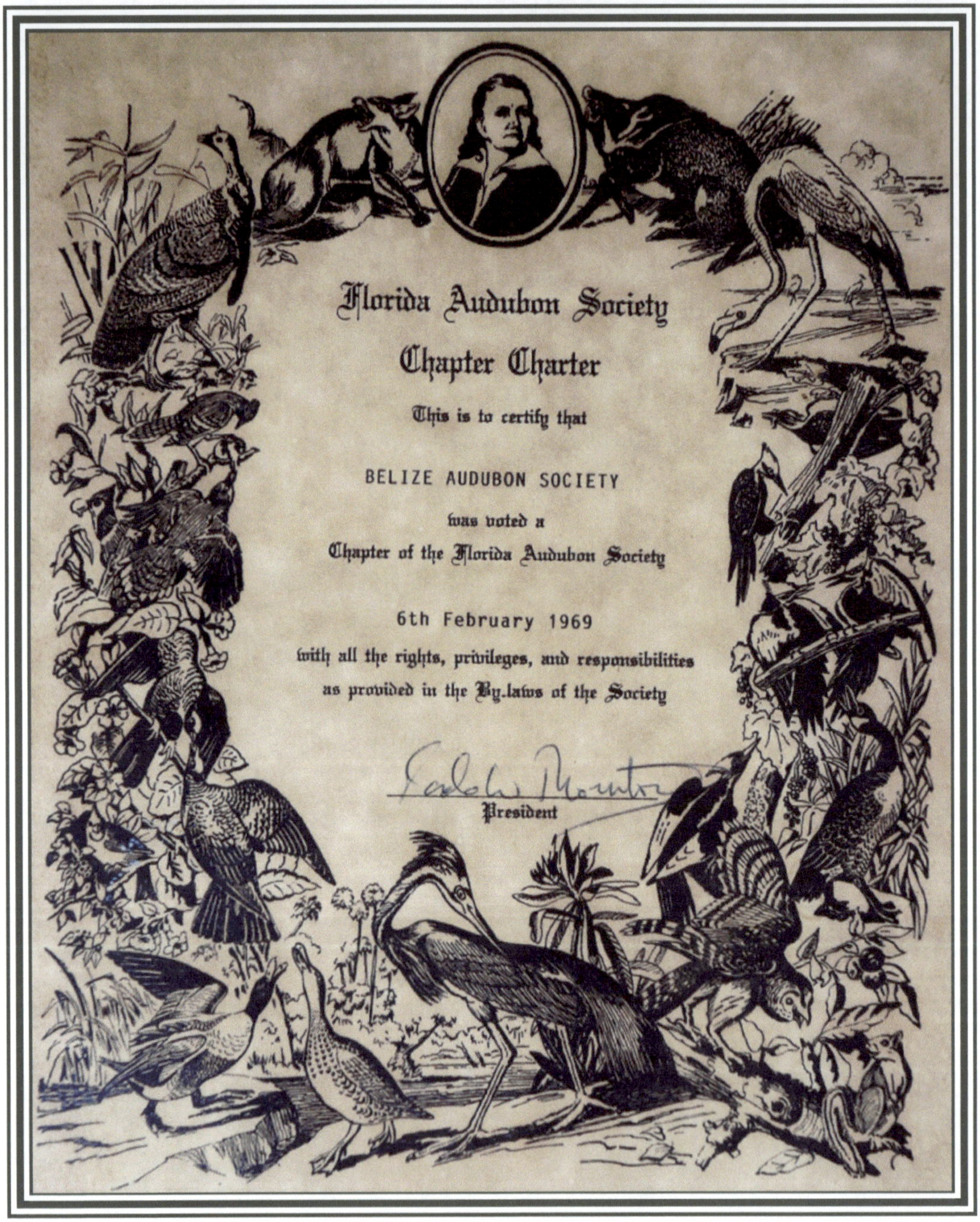

Florida Audubon Society

Chapter Charter

This is to certify that

BELIZE AUDUBON SOCIETY

was voted a

Chapter of the Florida Audubon Society

6th February 1969

with all the rights, privileges, and responsibilities
as provided in the By-laws of the Society

President

Chapter One

Formation and Growth of the Belize Audubon Society

Long before the initiation of the Belize Audubon Society there was already an environmental consciousness in the developing nation of Belize. The family of the Premier, George Price, and others were interested in wildlife and its conservation. When Dora Weyer moved to Belize, she made friends with Meg Craig, Lydia and Jim Waight, and other people who were interested in wildlife and joined her on bird-watching trips. As a member of the Florida Audubon Society Dora knew that the Executive Director, C. Russell Mason (Russ), was interested in starting foreign chapters. He arranged a tour for members of the Florida Audubon Society with the intention of getting a chapter started. A meeting was held on February 6, 1969, at the Premier's Lodge.

"Our group here wants to organize a local chapter of the Florida Audubon Society, to be called the Belize Audubon Society, for the purpose both of learning more about our wildlife ourselves and of preserving it as part of our national heritage," said James Waight as he chaired the meeting.

First BAS Officers and Directors	
President	James A. Waight
First Vice President	Dora Weyer
Second Vice President	Louis Lindo
Secretary	Lydia Waight
Treasurer	Albert S. Grant
Directors:	
Fr. Leo Weber, S.J.	Colonel Martin C. Hastings
W. Ford Young	Albert Staine
Norman Staines	Henry Fairweather
Magnus Halcrow	Meg Craig
Fr. Charles Woods,S.J.	Ronald Clark
Gil Rosado	Winston Miller

The Florida Audubon Society was interested in bringing Natural History tour groups to enjoy the wildlife of Belize and offered assistance in the preservation of birds and mammals in Belize.

At this meeting a decision was reached to go ahead with the formation of the Belize Audubon Society as a chapter of the Florida Audubon Society and temporary officers and directors were elected. These officers and directors were instructed to nominate a slate of officers and directors to be considered during an inaugural meeting of the Society to be held on a date fixed by the temporary board. Fifty-five persons became founding members of the Society. The inaugural meeting of the members and friends of the Society was held at the Bliss Institute on March 7, 1969, and the officers and directors were elected and six Standing Committees appointed.

First BAS Standing Committees	
Field Trips	W. Ford Young (head), Esther Pendergast, Col. M.C. Hastings
Programme	M. Halcrow (head) L.C. Balderamos, Jim Parker, W. Barmon, Peter Hill
Conservation	Louis Lindo (head), Winston Miller, Fr. Charles Woods, S.J., Paul Szabo, Henry Fairweather, Ronald Clark
Membership	Albert S. Grant (head), Albert Staine, Jim Parker, Hyacinth Waight
Youth Education	Fr. Leo Weber, S.J. (head), Meg Craig, Ann Carter, Norman Staines, Gilda Wagner, Kathleen Hope, Elvia Verde
Sanctuary	Dora Weyer (head), Premier George Price, Dr. Peter Schmidt, Gil Rosado, Fr. Len Dieckman,S.J., Jorge Verde

An Introduction to the Founders

Dora Weyer was the moving spirit behind the formation meeting. She and her husband had lived in West Africa where he worked with the Firestone Rubber Company. When he retired, they wanted to live in Belize and came here looking for property. Mr. Weyer died very suddenly before they could complete the move, but Dora and her daughter Diane settled here. Dora was an ornithologist, herpetologist and a wildlife person all around. She was very interested in the wildlife of Belize, especially in our birds.

BAS Archives

Mrs. Dora Weyer (1918–2007)

W. (Walker) Ford Young was born in Eugene, Oregon. He came to Belize in the 1950s with Gulf Oil Corporation after working in Nicaragua and Colombia. After retiring from Gulf Oil, he went into the real estate business, forming his own company in Belize.

James Waight was born in Belize City. In 1969 he had retired from his position as Surveyor General of Belize. He served the Society as President for 17 years.

Lydia Waight and **Alice M. (Meg) Craig** are sisters who were born in Belize. They grew up in a family that was greatly interested in

Lydia and James A. Waight

wildlife. Their brother, George Price, was at that time Premier of Belize. Lydia served the Society as Secretary for 27 years. Meg served the Society as Vice President. Her son Walter (Micky) was the first Executive Director of BAS, son David served as Vice President and President of BAS, and son Peter is one of Belize's best birders.

Colonel Martin C. Hastings, D. S. O., was the officer in command of the British forces in Belize. He had been in Kenya and was interested in wildlife conservation. Col. Hastings was most helpful in providing transportation for field trips and was able to supply maps for proposed reserves. In 1971 Col. Hastings moved to Devon, England, but he kept up his membership.

Esther Pendergast and her husband, David, an archeologist from the Royal Ontario Museum, both worked at Altun Ha. They were credited with naming Altun Ha, a translation of the local name, Rock Stone Pond, into Yucatec Maya. Herself a keen birder, Esther studied the birds of Altun Ha and added several new records for birds in Belize. Later she conducted tours to Belize along with D. H. Baldwin.

Magnus Halcrow was an Englishman who was the manager of the Reconstruction and Development Corporation that built Belmopan, the new capital. His wife, Molly, collected orchids and together they published the first work on the orchids of Belize. Mr. Halcrow was instrumental in the designation of Guanacaste Park as a protected area.

Russell Waters from Glasgow, Scotland, was the Chief Forest Officer. He met a tragic death on December 8, 1969, when the plane in which he was travelling crashed into the sea near the Haulover Bridge.

BAS Archives

(*from left*) James Waight, Shelly Matus, the guide, Michael Heusner (*standing*), Lydia Waight (*center*), Bill Hasse, Lou Nicolait, and Leo Balderamos. Photo by W. Ford Young

Louis Lindo was the Chief Game Warden at the time of the formation of the Society and succeeded Russell Waters as the Chief Forest Officer at the end of that year.

Fr. Leo Weber, Fr. Charles Woods, and Fr. Len Dieckman were Jesuit priests teaching at St. John's College in Belize City.

Dr. Peter Schmidt from Germany was the Archeological Commissioner at the time of the formation of the Society.

Jorge Verde lived in Sarteneja and had a tourist business there. He was the first to report on wood stork nesting in the Corozal District and on the slaughter of young birds in the nest. His wife Elvia was a teacher.

Ronald Clark was Secretary of the Tourist Board and later became Permanent Secretary in the Ministry of Trade and Industry.

Henry Fairweather, a surveyor, was Head of the Housing and Planning Department of the Government of Belize.

Winston Miller was Fisheries Biologist and later Fisheries Administrator.

Gil Rosado was Fisheries Officer in the Fisheries Laboratory of the Ministry of Trade and Industry.

Albert S. Grant was Chief Draughtsman in the Survey Department.

Albert Staine was Director of Public Prosecutions at the time of the formation of the Society. He later became Chief Justice and was knighted.

BAS Archives

Birding on a field trip to Guanacaste - Meg Craig is second from the left looking through binoculars; next is Sheree Smiling (later Mrs. Peter Craig); Philip Balderamos is just to the right of the Land Rover; and Dora Weyer is in the front on the right.

Norman Staines was Executive Officer of the Scout Association of Belize.

Jim Parker was a volunteer with the U.S. Peace Corps, teaching biology in Belize.

W. Barmon was the U. S. Vice Consul.

Peter Hill was Veterinary Officer at Central Farm

Paul Szabo was a Canadian surveyor working with the Survey Department.

Hyacinth Waight was Information Department Secretary.

Ann and Bolon Carter were Papal volunteers working in Belize.

Kathleen Hope was a teacher.

Field Trips

W. Ford Young

Soon after the establishment of the Belize Audubon Society, the directors felt that it would be in the interest of the members of the Society to organize birding field trips on a regular basis. Ford Young volunteered to undertake the organization and conducting of trips on one Sunday of each month.

Ford was in the real estate business and had an open back Land Rover so he was able to provide transportation for the groups. Other members often brought their own vehicles and sometimes a participant would make a small donation to help cover the cost of gasoline. When the Land Rover got too old for this work, Ford got a Chevrolet pickup and installed upholstered seats and a roof with canvas

BAS Archives

Field trip to Bermudian Landing

curtains in the back. This was appropriately labelled the "Audubon Birder."

Ford enlisted the services of Meg Craig to accompany him on all trips and act as a recorder. In addition to being a project recorder, Meg is also a knowledgeable self-trained botanist and was able to contribute greatly by identifying various plants encountered on the trips.

Dora Weyer was an excellent ornithologist and was most helpful in identifying unusual species when she could participate.

Trips were taken each month and were either half or full day trips. The first trip was a half day to the Mayan ruins of Altun Ha, some 32 miles from Belize City just off the old Northern Highway. Thirty-seven species of birds were identified. The number of participants on the trips ranged from about 6 to 30. It was most gratifying to note that a number of students were regular participants. It was not unusual for birders from abroad, who were visiting Belize, to join the trips. Summaries of each Sunday's trip were included in the monthly bulletin published by Mrs. Lydia Waight, first BAS Secretary.

The Early Years

The Society met every month and members took charge of the programme. There were talks, lectures, slide presentations and movies by members and others sharing their expertise. Business meetings usually followed the feature, which was advertised as being open to the public.

Table 1.3 BAS Programmes in 1969		
March 7	Dora Weyer	Birds of Belize City
April 17	Magnus Halcrow	Orchids of Belize
May 22	W. Ford Young	Physical Geography and Geology of Belize
June 19	Meg Craig	Tropical Plants
July 24	Louis Lindo	Forest Environment
August 21	Peter Hill	Movies of Wasps, Butterflies, and Snakes
September 12	Dr. Andre Clewell	Use of Fire in Forest Management
September 18	Gil Rosado	Our Fishery Problems
October 23	Dora Weyer	Snakes of Belize (with live specimens)
November 20	Peter Hill	Movies of Birds, Mammals, and Insects

First Annual General Meeting

The first BAS Annual General Meeting was held on Thursday, February 12, 1970, in St. Catherine's Convent Auditorium. Following the report by the President of the accomplishments of first year, plans were made for the second year. A vigorous membership drive was suggested because *"we feel that our impact would be greater in proportion to a larger membership. Requests to Government will have a greater influence and our influence on the attitude of the people of this country towards their wildlife and natural resources will increase with a larger membership."* (*BAS Bulletin,* February, 1970)

Other plans included an educational programme for police on the wildlife laws; efforts to protect the Jabiru, jaguars, ocelots, and Morlet's crocodile; creation of the Half Moon Caye Natural Monument; setting aside Cayo Rosario and Cayos Pajaros as bird sanctuaries for breeding colonies of Roseate Spoonbills, Reddish Egrets and White Ibis, environmental education for children; a programme of requesting private landowners to set their land aside as private reserves; and creation of the Crooked Tree Wildlife Sanctuary. Over the next few years these were all accomplished, as is documented in the following chapters on Environmental Education, Advocacy, Protected Areas Management, and Research.

Becoming Independent

Lydia Waight

At first we were members of Florida Audubon Society. Membership fees were paid in U.S. dollars and sent to Florida Audubon who would remit part to BAS. We had to adopt the charter of Florida Audubon as our constitution because we were a chapter, but made our own Belizean bye-laws. On June 25, 1969, the BAS Board met in the Forest Department Belize City office to consider the proposed bye-laws. By August 18 the bye-laws were completed and sent to Russ Mason at the Florida Audubon Society.

Florida Audubon initiated the severance of the chapter relationship because it was difficult for them to handle these foreign chapters. Several meetings were held to form a Pan-American group of independent Audubon societies. The new organization was called *Amigos de la Naturaleza* (Friends of Nature). *Amigos* only lasted a couple of years, but it did facilitate the process of BAS becoming an independent Society in 1973.

BAS Logos

FIRST BAS LOGO

Diane Weyer, daughter of BAS founding member Dora Weyer, designed the BAS logo, which records a common occurrence at Half Moon Caye. A Magnificent Frigatebird, locally known as "man-o-war," in hot pursuit of a Red-footed Booby returning from a long day of fishing far out at sea, seeks to make the booby drop its catch. The frigate will then skilfully seize the falling morsel, leaving the hapless booby to fly home with nothing to feed its chick. However, frigates and boobies have coexisted at Half Moon Caye for a very long time.

BAS Secretary Judy Lumb presented an award to Elise Craig in gratitude for her design: the beautiful newly upated logo.

UPDATED BAS LOGO

In 2005 Elise A. Craig donated her time and expertise to create an updated BAS logo. She described the new logo in the branding document. "While keeping these same easily identifiable features of the original logo, an important change is made to the encircling feature. The circle which held the birds "captive" is transformed into two open swirls in the shape of a crescent half moon. Adding this half moon symbol to the logo further strengthens its significance, conjuring up the image of the exquisite island, the very first protected area managed by BAS, where these birds roost in raucous, uneasy coexistence. In the new logo, the Magnificent Frigatebird with its distinctive scissor tail appears to fly across and in front of the crescent moon, adding much needed energy to the scene."

The First Fifteen Years

This letter from Eugene Eisenmann, Vice President of the International Council for Bird Preservation—Pan-American Section, provides a nice summary of the first five years of the Belize Audubon Society.

> *I peruse with interest and admiration your Bulletin, which has appeared with regularity. The President's report of 1974 is one of which you may well be proud. Not only have you issued the Bulletin with regularity and held frequent field trips, organized lectures and other educational activities, but you have been exceptionally effective in persuading the government of your country to set aside national parks and reserves. You are fortunate in having a government whose leaders are aware of the importance of reserving wild habitats and fauna for aesthetic, patriotic, as well as touristic and economic reasons. Your natural history postage stamps are among the most beautiful of any issued and it is of interest to know that a new issue of bird stamps and coins containing birds will soon appear. Let me congratulate the officers and members of the Belize Audubon Society for having made this relatively small organization one of the most vigorous conservation and ornithological societies in tropical America. (Letter from Eugene Eisenmann, 3 June 1975)*

BAS President James Waight responded:

> *"We are very fortunate in having an active and interested group as members of our Board and I know that your recognition of the results of their efforts will be greatly appreciated. A great deal of credit must be given to Mrs. Lydia Waight, our very capable and efficient secretary who edits the monthly Bulletin."*

During the next ten years, from 1975 through 1984, the structure and activities of BAS remained essentially the same. The business of the Society was carried out by monthly meetings of the Board of Directors and the Annual General Meeting of the entire Society. Lectures by visiting scholars were held whenever opportunities arose. The work of the Society—environmental education, management of protected areas, and advocacy—was done by volunteers with donated equipment and materials. BAS remained the only environmental organization in Belize.

A Time of Change

The end of 1984 brought many changes to BAS. It was the beginning of grant-funded projects to support the management of protected areas. Funds were obtained to open an office and hire staff. For the first time in its history as an independent nation, the Government of Belize changed when the opposition party won the election in December of 1984. Two proposed development projects for Belize brought international controversy and attention, along with conflicts within the Society. Other environmental organizations were formed. In short, everything changed.

When Government changed in 1984, concerns were raised whether the new government would support Audubon. The invitation to the sixteenth BAS Annual General Meeting (AGM) was issued, along with the slate of officers proposed by the nominating committee. But many members raised concerns about the slate because most of the officers nominated were supporting the Crane Game Ranch project (*see Chapter 2 Advocacy*).

BAS Archives

Dr. Victor Gonzalez presides over the 19th BAS Annual General Meeting in 1988

The AGM was held as scheduled on February 19, 1985, but was adjourned until March 12, when a new slate was presented and elected. To resolve this crisis, James Waight served as President for one more year. The next year Dr. Victor Gonzalez was elected BAS President.

Some of the American expatriate members of the BAS Board formed a new society, the Belize Heritage Society. It was to be an umbrella organization that could respond quickly to environmental crises and to fund-raising opportunities, but it only lasted a few years.

Despite the expressed concerns, BAS worked well with the new government. Hon. Dean Lindo, the new Minister of Natural Resources, was very supportive of conservation in general. He signed the documents for establishment of several reserves, including Hol Chan Marine Reserve, Guanacaste National Park, Blue Hole National Park, Tapir Mountain Nature Reserve (then Society Hall Nature Reserve) and Cockscomb Basin Wildlife Sanctuary.

(*from left*) Therese Bowman (Rath), Curtis Freese of the World Wildlife Fund-U.S., Lydia Waight, James Waight, and James Baird of the Massachusetts Audubon Society in 1987

BAS Administrative Staff

By the end of 1984, BAS had received several grants to support the management of protected areas in Belize, but until the offer from the Massachusetts Audubon Society (MAS), none had included any support for general administration.

The BAS Office at 49 Southern Foreshore opened January 2, 1985, and Walter (Micky) Craig was employed as Executive Director. A year later MAS announced a five-year commitment of increased support, allowing for a full-time Executive Director.

In April of 1987 a group from MAS, including four MAS Board members, visited Belize and reiterated their support for BAS. They said that this support was to help BAS manage the established reserves.

But two months later a MAS representative visited a BAS Board Meeting and expressed concern about BAS progress. By the Fall of 1987 the level of funding from the MAS had been reduced and the Executive Director position reduced to half-time. BAS sent MAS a 19-page report that showed details on park management, 216 percent increase in number of members and 367 percent increase in income from memberships, and reported $30,221.18 in the endowment fund.

But this was not satisfactory progress from MAS' point of view. BAS had been proceeding on the basis of their objectives stated in the original proposal letter, but the MAS expected long-range planning. Following the January 19, 1989, BAS Board of Directors meeting, MAS sent a document describing MAS' long-range planning process for

BAS to use as a guide. However, the funding had already been withdrawn. A donation of $1,000 from the Regent Insurance Company Ltd. prevented the closure of the office.

In March of the same year MAS agreed to provide $750 US a month to allow BAS to keep an office open with clerical support, but without an Executive Director. In the meantime, MAS announced the formation of a new organization, Programme for Belize, which would manage the lands that MAS was prepared to purchase in Belize.

MAS made another contribution to BAS Administration. Through Programme for Belize, they provided the funds for the salary of a trainee, Dolores Godfrey, who began on February 1, 1990, and became BAS' second Executive Director.

BAS Archives

BAS' first Executive Director, Micky Craig

BAS Archives

BAS Executive Directors: (*above*) Dolores Godfrey accepts a donation check from James Beveridge. (*below*) Virginia Vasquez

BAS Archives

Other Executive Directors include:

Virginia Vasquez (1993–1995)

Ted Castillo (1995–1997)

Osmany Salas (1997–2001)

Valdemar Andrade (2001–2004)

Anna Dominguez Hoare (2004–2010)

Amanda Burgos Acosta (2010–present)

Wild Wings Foundation provided the funds for a Protected Areas Manager, Osmany Salas, who began work in August of 1990. A joint project with the Coastal Zone Management Project provided funds for an Environmental Educator and June Neal was appointed in 1991. In 1993 Christine Anthony was appointed as BAS' first Accountant; Pedro Perez became the first Membership and Marketing Coordinator 1997; and Valdemar Andrade was appointed the first Advocacy Coordinator in 1998.

BAS Archives

BAS Executive Directors: (*above*) Valdemar Andrade with BAS founder Lydia Waight; (*right, top*) Osmany Salas on the new boardwalk at Crooked Tree; (*right, bottom*) Anna D. Hoare

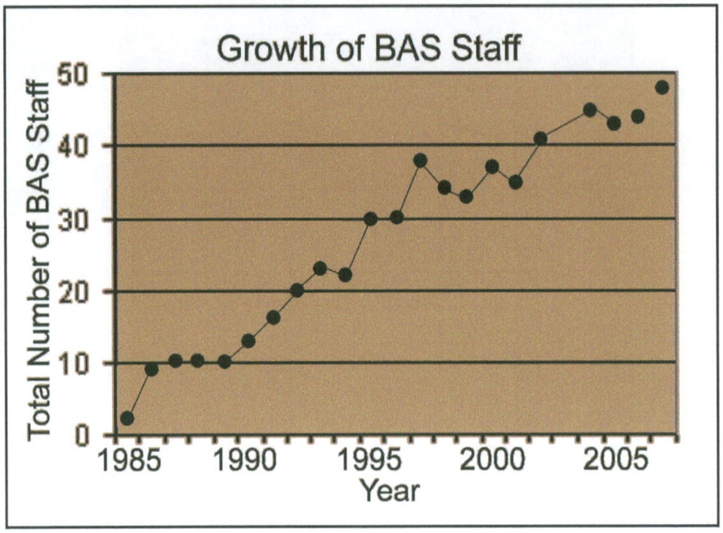

Growth of BAS Staff

BAS Executive Director Amanda Burgos Acosta began her tenure in 2010.

By 2000, BAS was managing six of Belize's protected areas and the work was more than one person could manage, so the Protected Areas Manager portfolio was divided among three positions. Julieanne Robinson was the first Marine Protected Areas Coordinator (later termed "Manager"); Michael Somerville was the first Terrestrial Coordinator (Cayo Parks); and Valdemar Andrade was the first Co-management Coordinator (Cockscomb Basin and Crooked Tree Wildlife Sanctuaries). At the same time Christopher Hecker was appointed as the first Natural Resources Data Manager (later Research Manager) and Jamie Lee Chanona was the first Marine Biologist. Leanne McCarthy was the first Marketing and Development Coordinator. Volunteer Dirk Francisco became the Public Participation and Publicity Coordinator.

As BAS developed a closer relationship with the communities, a Community Liaison Officer was added to the Staff. Heron Moreno became the first to hold this position in 2002.

When BAS built a gift shop, a Gift Shop Clerk was added with Colleen Hamilton in that position. In 2009 Kendra Griffith became the first Membership Coordinator.

The BAS Field Staff expanded along with the Office Staff. A complete list of BAS Staff, past and present, is found in the Chapter Seven. The graph (*above, left*) demonstrates the growth of the BAS Staff to the 48 creative, energetic conservationists who implemented the 40th BAS Anniversary celebration in 2009.

BAS Membership

BAS is a membership organization and membership is open to anyone concerned with preserving our natural heritage. The first year there were 55 members and that number had doubled by 1981. At the 30th anniversary in 1999, BAS membership totalled nearly 400 local members, and over 200 foreign associates and by the 40th anniversary there were 1211 members, 227 of whom were foreign associates.

Besides the essential support of the work of BAS, striving for a balance between people and the environment, BAS membership carries a number of benefits. Each year there are several special events for members: Summer Camps, Earth Day activities, James A. Waight Award Ceremony, Annual General Meeting, and many birding and other outings.

In 2008 BAS held its first Membership Convention on April 26th–27th. Approximately 109 members attended the convention to experience how the vision and mission of BAS are realized through our conservation efforts. On the first day each BAS Programme Manager creatively presented their programme area to the members present, sharing facts about the work they do and why members' support is necessary to continue our work. The Kid's Corner engaged children in educational and environmental videos, puzzles and art designs. The children presented a song based on "We Are the World," and a monologue, "It all started with the Boobies." On the second day members were given a first-hand experience of the work of BAS Staff at Half Moon Caye and Blue Hole Natural Monuments.

Day 1 of the BAS Membership Convention

Day 1 of the BAS Membership Convention 2008
(*above left*) Ernesto Vasquez as Master of Ceremonies
(*below left*) Children in Kid's Corner
(*above right*) Presentation to the Membership Convention by Yassiri Cruz, a volunteer from Sarteneja Village.

BAS Membership Convention 2008

BAS Organization

The governing body of BAS is its Board of Directors, consisting of six officers, (president, two vice-presidents, two secretaries and treasurer), the immediate past-president, and a maximum of eight other directors. The Board is elected at the Annual General Meeting for staggered two-year terms.

In 1972, when BAS wanted to buy the area that became Guanacaste Park, their attorney told them that a membership club could not own property. So, they set up a Board of Trustees, including Ford Young, James Waight and Meg Craig. That continued until 1990 when BAS became a limited liability company and the Board of Directors could represent the Society.

In 1997 BAS created another Board of Trustees to manage the BAS Endowment Fund. In 2006 the functions of the Board of Trustees were taken over by the BAS Board Finance Committee. Awards were given to the members who had served the Society so well and the Board of Trustees was discontinued.

Day 2 BAS Membership Convention, April 27th, 2008, at Half Moon Caye

Annual General Meetings

As a membership organization, BAS members gather for the Annual General Meeting (AGM). Until 2006, AGMs were held in the evening at hotels in Belize City. Some guest speakers at AGMs are listed in Table 1.4 .

At the 30th AGM the membership gave the Board of Directors the task of developing a Code of Conduct, which was presented to the 31st AGM. One section states that BAS Board members cannot hold political office, including being on a Village Council. Concern was raised that would eliminate people from smaller communities as most active people in a village serve on their Council. The Board polled the membership through the *BAS Newsletter* and decided that even being on a Village Council can create conflicts of interest, so the Code of Conduct was approved with no changes.

Beginning with the 37th Annual General Meeting in 2006, the format of the BAS AGM was changed to Saturdays at BAS-managed protected areas. We discontinued the practice of having Guest Speakers. The business meeting was held in the morning and activities were planned in the afternoon to highlight that protected area.

Table 1.4 Guest Speakers at BAS AGMs

AGM	Speaker	***Position/Organization***
21st	Dr. Rob Horwich	Community Baboon Sanctuary
22nd	Janet Gibson	BAS President's Report
"	Osmany Salas	Protected Areas Manager Report
23rd	Officers and Staff	Staff and Committee Reports
24th	Lindsay Belisle	Ministry of Natural Resources
25th	Mark Nolan	NARMAP Programme
26th	Robert Dakan	USAID
27th	Nestor Windevoxhel	IUCN Regional Mesoamerica
"	Julian Cho	Toledo Maya Cultural Council
28th	Marcel van Opstal	European Commission
29th	Osmany Salas	BAS Annual Report
30th	Hon. Ralph Fonseca	Ministry of Budget and Planning
"	Dr. Alejandro Grajal	National Audubon Society
31st	Osmany Salas	BAS Annual Report
32nd	Dr. Teny Topalian	UNESCO
33rd	Valdemar Andrade	BAS Annual Report
34th	Hon. John Briceño	Minister of Natural Resources
35th	Osmany Salas	Chief Forest Officer
36th	Sylvana Woods	Environmental Message

31st BAS AGM May 30, 2000, Radisson Fort George Hotel (*sitting, from left*) Judy Lumb, Lisel Alamilla, Jose Garcia (*standing*) Mike Heusner, Reynaldo Guerrero, Ambrose Tillett, Amin Bedran, John Balderamos, David Craig, Earl Green (*not pictured*) Humberto Paredes, Anselmo Castenada, Wilmot Simmons, Magali Marin and Michael Singh

34th Annual General Meeting

(*left*) Christine Anthony, Nancy Tillett, and Yesenia Leiva at the AGM Registration table.
(*right*) The crowd at the 34 BAS AGM held May 7th, 2003, at the Radisson Fort George Hotel.

BAS Board of Directors elected at the 34th AGM
(*back row, from left*) Earl Green, John Balderamos, Carlos Santos, Humberto Paredes, Mark Hulse, (*front row*) Patrick Scott, Sr., Amin Bedran, Anna D. Hoare, Mike Heusner, Judy Lumb

35th Annual General Meeting

The 35th BAS AGM was held on May 24th, 2004, at the Radisson Fort George Hotel (*top*). The Keynote Speaker was Chief Forest Officer Osmany Salas (*center*), formerly BAS Executive Director. The new Board: (*bottom, back row from left*) Jude Lizama, David Craig, Patrick Scott, Sr., Earl Green, Mike Heusner, Carlton Young (*front row*) Nicole Auil, Lascelle Tillett, John Balderamos, Anna Rossington, Judy Lumb, Linda Searle

The 37th AGM was held on May 27th, 2006, at Guanacaste National Park where activities included birding, kayaking, swimming and hiking. The 38th AGM was held at St. Herman's Blue Hole National Park on May 26th, 2007. Activities included birding, swimming and dancing to marimba music.

A wonderful plan to hold the 39th AGM in the Cockscomb Basin Wildlife Sanctuary (CBWS) was thwarted when two tropical storms collided over Belize: Tropical Storm Alma, the first Pacific storm of the season, and Tropical Storm Arthur, the first Atlantic storm of the season. A major bridge on the main Southern Highway that provides access to the southern part of the country, including CBWS, was completely washed away.

The 39th AGM on 31st May 2008 was moved to the Belize Biltmore Plaza Hotel in Belize City. While everyone was disappointed because all had been looking forward to going to Cockscomb, the BAS Staff did a great job of adjusting the plan so the AGM was as successful as always, including Maya dancing. The next year that wonderful plan from the year before was finally executed as the 40th AGM, held in the Cockscomb Basin Wildlife Sanctuary.

37th Annual General Meeting

The 37th BAS AGM was held at Guanacaste National Park on May 27th, 2006. (*above*) Executive Director Anna D. Hoare asked that the Staff stand behind her as she presented the Annual Report. (*below*) Orlando Jimenez at the barbecue. (*right*) David Craig on top of a branch cut off the famous Guanacaste tree to avoid the danger of it falling.

Judy Lumb

38th Annual General Meeting

The 38th BAS AGM was held at St.Herman's Blue Hole National Park on May 26, 2007. (*top*) The Staff stands behind Mrs. Anna D.Hoare, the Executive Director as she presents the Annual Report. After the meeting in the morning, all members enjoyed a delicious barbecue and various actvities such as birding (*above left*) or swimming in the Blue Hole (*above right*). The newly elected Board: (*from left*) Nicole Auil, Lascelle Tillett, Linda Searle, Carlos Santos, Mario Teul, Earl Green, Jude Lizama, Anicasio Garcia, Carlos Matus.

39th Annual General Meeting

(*top left*) Peace Corps Volunteer Scott Schmidt with children at 39th AGM, which was moved from Cockscomb to Belize City, along with Maya Mopan marimba players (*bottom left*); (*top right*) BAS Director Lascelle Tillett and his wife, Sarita; (*bottom right*) Maya Mopan dancers at 39th AGM at the Belize Biltmore Plaza, Belize City

40th Annual General Meeting

Mike Heusner

Mike Heusner

(*top left*) BAS members at 40th AGM held in Cockscomb Basin Wildlife Sanctuary (CBWS); (*bottom left*) BAS Executive Director Anna D. Hoare, BAS President Earl Green, BAS Treasurer Mark Hulse; (*top, right*) New Visitor Center at CBWS opened by President Green and Rosita Pop, President of the Maya Center Women's Group; (*center right*) Dominque Lizama and Khaidia Mossiah serving the barbeque dinner; (*bottom right*) CBWS Park Director Nicacio Coc giving the Vote of Thanks

Cooperation with the Government

Lydia Waight

BAS has had excellent cooperation from the Government of Belize throughout its history. Gift subscriptions to wildlife and conservation magazines were sent to the Chief Forest Officer and other Government officials. This provided the conservation point of view to those in the highest positions.

BAS lobbied the Government for the creation of the position of Conservation Officer. Lucilo Sosa Briceño was appointed the first Conservation Officer in June of 1971. We invited Mr. Sosa to attend our meetings and requested that the Minister of Trade and Industry appoint a liaison officer who would attend BAS Board Meetings. Lionel Gentle of the Forest Service was appointed Liaison Officer.

Government consulted us on matters relating to the environment, wildlife protection and developmental proposals. As our requests for particular reserves were granted, our involvement in the management was expected.

This arrangement was first formalized in 1984 by a letter from James V. Hyde, Permanent Secretary, Ministry of Natural Resources. BAS was authorized to "work along with the Forest Department in the management of protected areas, draft and implement management plans, and accept contributions and payment for services such as the rental of facilities. Responsibilities were further delineated in 1987 by a letter from R. H. Neal, Permanent Secretary, Ministry of Agriculture, Forestry and Fisheries.

In 1995 BAS signed a five-year agreement that defined the responsibilities for management of six protected areas, Half Moon Caye Natural Monument, Crooked Tree and Cockscomb Basin Wildlife Sanctuaries, Guanacaste and Blue Hole National Parks and Tapir Mountain Nature Reserve.

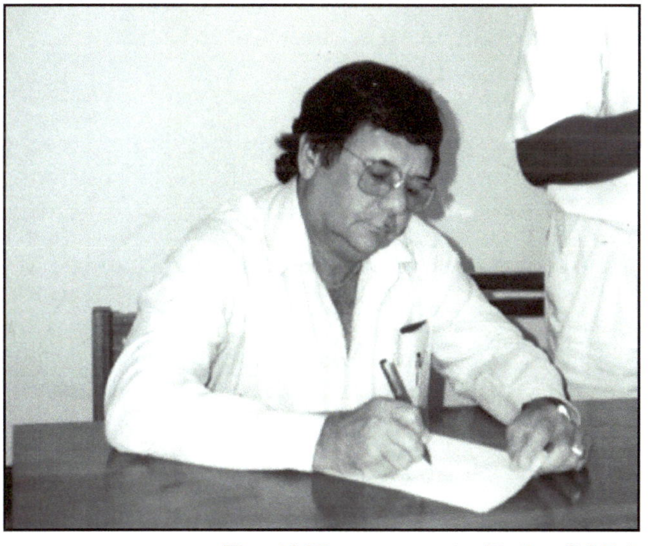

BAS Archives

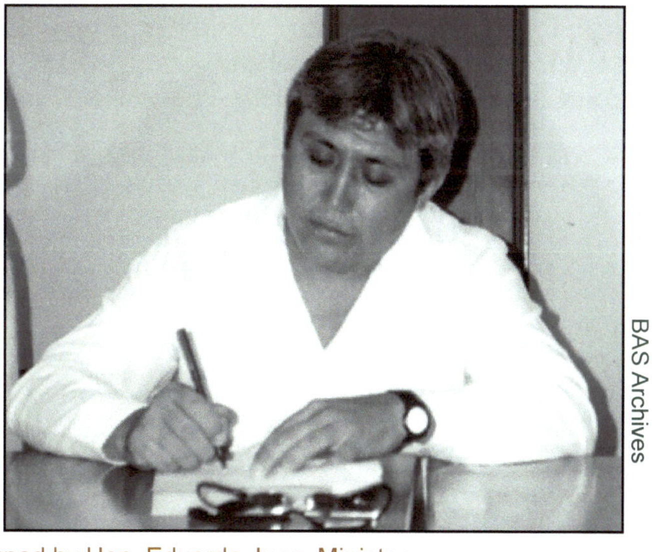

BAS Archives

The 1995 agreement with the GOB is signed by Hon. Eduardo Juan, Minister of Natural Resources (*left*), and Jose (Pepe) Garcia, President of BAS (*right*).

It was agreed that Government, the Forest Department and the Society would jointly manage and develop the six protected areas for a period of five years, renewable for a similar period. The parties were to jointly formulate management plans with goals, objectives, standards, methods of implementation, priorities, budget, personnel requirements, target dates, and assessment methods for monitoring. Provisions for transition periods and mechanisms for resolving conflicts were included.

In this agreement, the GOB was responsible for legal Statutory Instruments and security, as well as handling applications for scientific research permits in consultation with BAS. The GOB was to assist BAS with tax exemptions and authorized the Society to collect entrance fees.

BAS was authorized to obtain grants, loans, subsidies, endowment and trust funds to finance the management of these protected areas. BAS had responsibility for day-to-day management, implementation of all aspects of visitor usage and daily maintenance, and public education campaigns. BAS had first right of refusal for all recreation-related concessions with the profits for use by BAS.

An updated agreement between BAS and the GOB was signed on February 6, 1999 by BAS, the Chief Forest Officer and the Minister of Natural Resources. Two new areas were added to those managed by BAS—Blue Hole and Victoria Peak Natural Monuments.

On August 31st, 2004, a ten-year agreement was signed between the Forest Department and BAS. The ten-year tenure allows BAS to complete long-range planning for protected areas management. This agreement included Actun Tunichil Muknal Natural Monument, which is managed cooperatively with the Archaeology Department.

BAS Workshops and Retreats

Park Management Workshop

From March 30 to 31, 1989, a Peace Corps-sponsored workshop was held for BAS staff and committee chairpersons on park management. The Society was getting quite a number of requests from scientists wanting to research in our protected areas. The purpose of this workshop was to develop recommendations and policies on research priorities and guidelines, use of parks by visitors, and a training programme for park personnel.

A Vision for the 90s

Through a grant from the World Wildlife Fund–US, a Strategic Management Workshop was held on April 29, 1990, at Bridgewater, Burrell Boom. Through a series of brainstorming sessions, the BAS officers and staff evaluated their protected areas management and made recommendations for the future institutional development of the Society. The activities of the Society then included:

- Management of reserves
- Creation of new protected areas
- Influencing resource use
- Environmental education
- Improving our image and public relations
- Collection and dissemination of environmental information

The Board planned to support this work through contributions from local companies, foreign conservation organizations, but committed to "explore means of becoming as self-sufficient as possible."

The Board also committed themselves to increase involvement of our membership, to conduct a membership drive, and to cooperate with the other local non-governmental organizations to achieve our common goal of sustainable development for Belize.

BAS Board/Staff Retreats

The first joint BAS Board/Staff Retreat was held in 1991 at Trinidad Farm and facilitated by Belize Enterprise for Sustained Technology (BEST). The purpose was to review BAS' current status and programmes in the light of the environmental issues of the 1990s. One product of this retreat was the BAS Mission Statement:

The Belize Audubon Society is a nonprofit, non-government organization dedicated to the promotion of the sustainable use and preservation of our natural resources in order to maintain a balance between people and the environment.

A follow-up Strategic Planning Workshop was held October 26 and 27, facilitated by Johannah Barry of the IUCN with Byron Swift, also of the IUCN, attending.

Participants around this table in 1991 are BEST Facilitator Carlos Santos, James Waight, an unidentified BEST Facilitator, John Link, Rudi Burgos, Shelley McCrary, Therese Rath, Osmany Salas, Bardi Riverol, David Craig, Lydia Waight, Luz Hunter, Magda Avery, Janet Gibson, Victor Gonzalez, Dolores Godfrey, and Philip Balderamos. (*photos pp 27 and 28 from BAS Archives*)

Participants in the same September, 1991, retreat at Trinidad Farm include (*above from left front*) John Link, Bardi Riverol, Rudi Burgos (*standing*), Magda Avery, Dolores Godfrey, (*below from left*) David Craig, Shelley McCrary, Osmany Salas, Pepe Garcia, Magda Avery, and Carlos Santos

The third BAS Retreat was held at Spanish Bay Resort in August of 1992. For the first time the park personnel were included. After two days of hard work, President Therese Rath said,

> *"What we have done here is to focus on one area in more detail as to what our programme should be and to discuss what other areas we should be focussing on."*

At the 1993 Board retreat, it was decided to re-focus the Society's activities from protected areas management to environmental education as the main vehicle to "promote the sustainable use of our natural resources" in keeping with our Mission Statement.

Then at the 1994 BAS Retreat it was decided to add a third objective, advocacy, to the other main activities of protected areas management and environmental education.

Additional BAS Retreats were held in 1997 at Crooked Tree Wildlife Sanctuary, in 1999 in Corozal Town, in 2000 at Lighthouse Reef Resort, in 2004 at Cockscomb Basin Wildlife Sanctuary, in 2006 at Cedar Cabins, in 2007 at Chaa Creek Resort, in 2008 at Cahal Pech, and in 2009 at Cockscomb Basin Wildlife Sanctuary.

BAS Programme Review and Assessment

The British Executive Service Overseas (BESO) provided a volunteer consultant, John Bishop, to review and assess BAS programmes. Bishop worked with BAS Executive Director and Programme Managers from July 20 to 23, 1998, and made the following recommendations:

- Petition the GOB to increase park entrance fees for foreign visitors
- Promote BAS membership in the National Parks and provide training for wardens in recruitment techniques
- Improve signposts and quality of interpretive literature in the parks
- Ascertain accurate membership recruitment and retention costs, review all subscription categories, and closely monitor retention rates
- Identify and train a network of regional volunteers
- Take advantage of the Society's 30th anniversary to organize fund-raising and public awareness events.

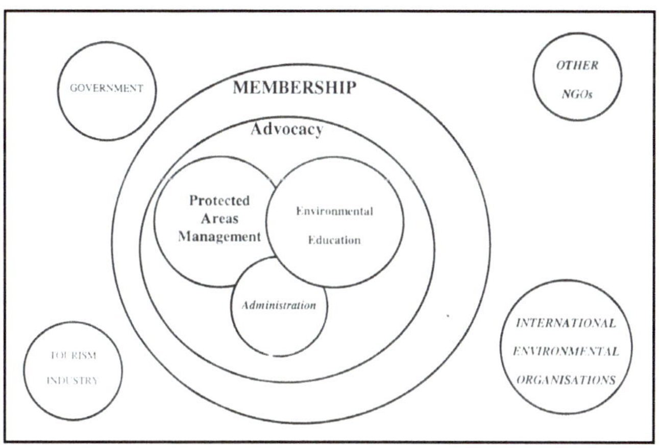

The image of BAS in the Strategic Plan for 1996–2001

Strategic Plans

Strategic Plan 1996-2001

At the 1994 annual BAS retreat at Rio Bravo all present agreed it was time to work on a long-term plan. "Belize Audubon Society to the Year 2001: Strategic Plan for 1996–2001" began with a list of global threats from the Convention on Biological Diversity and then considered Belize's environment in that context. A technique called "Strengths, Weaknesses, Opportunities, and Threats (SWOT)" analysis was used to assess BAS' greatest strengths and weaknesses, and opportunities for success. Within the programme areas of Advocacy, Education and Protected Areas Management, an ambitious plan was developed to address all of Belize's environmental problems. Knowing that plan would require considerable effort by many people, the goal was set to double the BAS membership every year for four years, arriving at 20,000 by 2001.

Strategic Plan 2002-2006

As the term of the first strategic plan grew to a close, BAS embarked on a strategic planning process supported by HIVOS and facilitated by Shaun Finnetty. "Towards the New Millennium: Strategic Management Plan 2002–2006" was completed in December 2001. A performance audit again using SWOT analysis generated the following, "BAS Current Performance Statement (2001)."

> *BAS is able to manage many protected areas with limited resources. The organization has grown from a small entity with limited resources to become a leader through its management, creativity and innovation of staff and the Board. It has risen to challenges in the past and is planning strategies to meet future ones. BAS is expected to set standards, which is a serious responsibility. The Society is under constant scrutiny and is reactive rather than proactive.*

It was felt that the Society was trying to do too much and was spread too thin reacting to urgent requests to solve all of Belize's environmental problems. To help the Society become more proactive and less reactive, the Strategic Management Plan 2002–2006 focussed on only two Programme Areas:

1) **Institutional Management and Development** included administration, finance, human resources, membership, public relations, systems development, organisational development, and infrastructure development.

2) **Integrated Resource Management** included protected areas management, education, research and advocacy. These programmes were focused on supporting the management of protected areas.

Strategic Plan 2009-2013

To begin BAS' third strategic planning cycle, the entire Staff and the Board attended a Strategic Planning Retreat at Chaa Creek Resort September 14–16, 2007, for which Oak Foundation provided some support. Drs. John and Kristen Hunt from the University of Idaho facilitated the first workshop with the help of Shaun Finnetty. This strategic plan included the views of key stakeholders and partners such as HIVOS, IUCN, Oak, WCS, PACT, TNC, our past presidents, and the membership as well.

The SWOT analysis noted BAS strengths:

The organization's extensive networking abilities and supportive institutional affiliations reflect the widespread recognition it enjoys and the many fruitful national and international partnerships formed over its years of existence. It currently benefits from having a talented and youthful staff imbued with a team-oriented approach to results-based work and performance.

Strategic Plan Staff Board Retreat, September 2007
(*top*) Elisa Ariola, Carol Williams, and John Arana
(*centre*) Carlos Santos, Scott Schmidt, and Dominique Lizama
(*bottom*) Khaidia Mossiah, Annette Reneau, Mario Bolon, Derick Hendy

The SWOT analysis indicated some weaknesses that became key targeted actions in the plan:

BAS is challenged by inadequate infrastructural capability in terms of workspace. ... Other factors such as the human resources management practices, the protected area management plan updates, overwhelmingly heavy programmatic workloads, technical support programme and financial planning are key to ensuring maximum effectiveness and so need to be given paramount importance. Secondarily, limited financial resources for managed protected areas, deficiencies in some technical areas as well as limited systematization and documentation of organisational learning strategies are also flagged concerns.

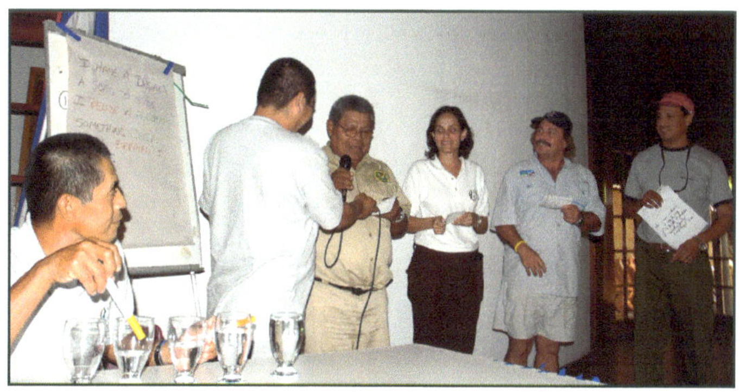

Staff-Board Strategic Planning Retreat 2007
(*above*) Roberto Pott; (*left top*) Kent Garcia, Albart Liconia, Linda Searle, Harvey Rosado, Ivis Chan; (*left centre*) Kent Garcia, Harvey Rosado, Albart Liconia, (*left bottom*) Marcelo Pau, Bonifacio Cal, Ignacio Pop, Nicole Auil, John Searle, and Roberto Pott

Staff-Board Strategic Planning Retreat 2007
(*front*) Khaidia Mossiah (*back from left*) Scott Schmidt, Annette Reneau, Rennie Jones, Orlando Jimenez, Mario Bolon, Derick Hendy, Felicia Leal

Opportunities include:

- building nontraditional private sector partnerships,
- building international educational partnerships,
- improving membership programs and services,
- engaging in public awareness initiatives,
- tapping into foreign philanthropic interests,
- revenue generation in educational tourism,
- increased political advocacy policy changes,
- mainstreaming environmental issues,
- enhanced Board involvement in public relations,
- fund-raising and strategic asset acquisition.

Constraints identified include:

- difficult national development dilemmas,
- increasing natural resource exploitation,
- a culture of undervaluing natural resources,
- an inadequate legal framework,
- lax regulatory enforcement,
- insufficient state allocations,
- weak community support,
- poor national educational outcomes,
- unemployment,
- unregulated mass tourism,
- unchecked political authority of government ministers,
- expanding petroleum exploration,
- natural disasters, and
- climate change.

The following Key Result Areas are the focus of the BAS 2008–2013 strategic plan:

1) Organisational Development
2) Protected Areas Management
3) Biodiversity Conservation
4) Environmental Education
5) Advocacy
6) Community
7) Research and Data Management

Biodiversity Conservation is a new activity area at BAS that recognizes ongoing work which for many years has focused "on the biodiversity conservation of birds and endangered species and incorporates species mapping and monitoring, scientific cooperation and data sharing as key strategic activities."

Financial Trials

The road has not been completely smooth for BAS. Three losses occurred within one year. On February 18, 1989, Ford Young's office was burglarized and, since his secretary was BAS' Treasurer, some $800 of BAS money was stolen. The next November a messenger forged $3,000 worth of BAS checks and left the country before it was discovered. Then on February 8, 1990, the upstairs flat had a plumbing leak which flooded the BAS Office, destroying many files. Considerable damage was done to posters and publications.

In 1995 BAS met with a financial crisis, the resolution of which instigated a series of measures to put the Society on a more solid financial footing. Until then BAS had borne the full financial responsibility for management of four protected areas, Blue Hole and Guanacaste National Parks, and Cockscomb and Crooked Tree Wildlife Sanctuaries. There were no entrance fees or contributions from the Government of Belize or the tourism industry.

BAS, led by President Therese Bowman Rath, approached the tourism industry, as well as the Ministries of Natural Resources and Tourism/Environment, and appraised them of the seriousness and urgency of the situation. The U.S. Agency for International Development pledged to temporarily assist with maintenance costs for six months and it was agreed that entrance fees and the Protected Areas Conservation Trust would be established.

BAS Staff-Board Retreat in December 2008
(*above*, *back row*) Transito Jimenez, Venancio Coc, Gillian Castro, Steve Tillett, Federico Tush; (*above, front row*) Nikisha Ack, Bonifacio Cal, Uriah Rhaburn; (*below*, *from left*) Erleon Garbutt, Shane Young, Ignacio Pop, Arlita Tillett, Carol Williams, Giovanni Martinez

Protected Areas Conservation Trust

For several years the Belize Audubon Society had been working along with the Government of Belize, other conservation NGOs and international conservation organizations, to develop a Protected Areas Conservation Trust, an innovative mechanism for sustainable support of the protected areas of Belize. The Act to establish a trust for the protection, conservation and enhancement of the natural and cultural resources of Belize, to establish a Trust Fund for the Trust and to establish a Board of Directors, was gazetted on September 30, 1995. Each visitor to Belize is charged a PACT fee of US$7.50.

Entrance Fees for National Parks

The Statutory Instrument for the charging of entrance fees for national parks and wildlife sanctuaries managed by the Belize Audubon Society was signed on May 20, 1995. Specific entry fees for Guanacaste and Blue Hole National Parks and for Crooked Tree and Cockscomb Basin Wildlife Sanctuaries for Belizean nationals and non-Belizeans were prescribed. Children under twelve years old are exempt from paying entry fees. The Government of Belize (GOB) authorizes the Society to collect fees and all fees collected shall be apportioned as follows:

PACT 20% for purposes of the trust
BAS 70% for management of the protected areas
GOB 10% for park infrastructure and security

The distribution of entrance fees was changed to 20 percent for PACT and 80 percent for BAS in 1999. By 2009 BAS received authorization to reinvest PACT's 20 percent of entrance fees into the protected areas. Ten percent of CBWS entrance fees were given to the Maya Centre Women's Group and ten percent of the CTWS entrance fees were given to Crooked Tree Village.

Financial Sustainability

As the BAS annual budget increased past BZ$1.9 million in 2002, it became clear that more detailed financial information was necessary for the Society to execute prudent financial management. An entire financial overhaul was done with upgraded software using international standards for nonprofit financial accounting and auditing. A Board Finance Committee was formed to oversee BAS operations.

By the end of 2004 BAS was in such difficult financial straits that it was not possible to give the Staff the traditional Christmas bonus. A few months later some inconsistencies in the financial records were reported. The Finance Committee investigated thoroughly and some changes were made in personnel and procedures, which were documented in a new *Financial Manual.* For the entire operation of the Society the *Policy and Operations Manual* was completed in 2008 after a lengthy process with all Staff and Board participating.

These improved financial procedures, prudent financial management, increases in entrance fees, and diversification of funding sources have brought the Society closer to the long-term goal of sustainability, which is indicated by the trend of decreasing dependence on grants (*graph below*).

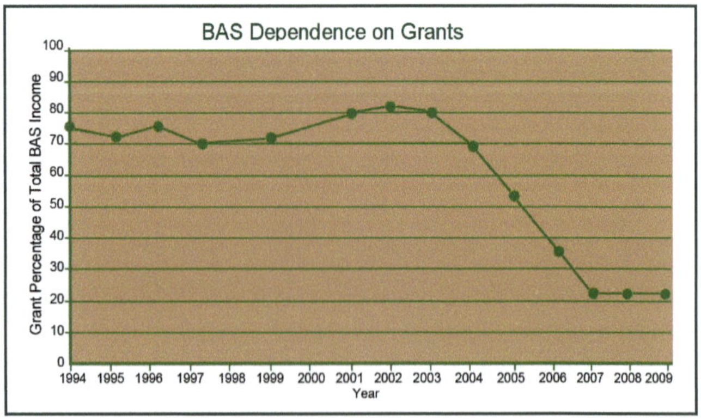

BAS Dependence on Grants

BAS Office

The first BAS Office, opened December 1, 1984, was located on the Southern Foreshore. Eventually that space became too small for the growing Society's activities and the owners wanted to reclaim the space. The BAS Office moved to 29 Regent Street on March 1, 1990.

In 1993 BAS was given the property at 12 Fort Street by the GOB, in appreciation of past and support of future BAS management of Belize's protected areas. At the 24th AGM, Permanent Secretary in the Ministry of Natural Resources Lindsay Belisle presented the title to BAS.

But then in late 1997 there were rumours that the BAS property at 12 Fort Street was to be acquired for a major Belize Port Authority development. In fact the Grant Fiat and Property Plan issued to the Belize Port Authority included the BAS property.

"The grant to the Belize Port Authority was issued in error and [the Commissioner of Lands and Surveys] is in the process of rectifying said Grant to omit Grant No. 306 of 1993 issued to the Belize Audubon Society. It is estimated that it will take about two (2) weeks to prepare and register the new Grant." (*Letter from Lindsay Belisle, Permanent Secretary, Ministry of Natural Resources, 21 January 1998*)

On April 14, 1998, an injunction order was issued from the Chief Justice which stopped the Minister from granting the Fort Street property to the Belize Port Authority until further notice. The situation was finally resolved on February 6, 1999, at the Opening of the BAS 30th Anniversary Celebrations, when Hon. John Briceño, Minister of Natural Resources, announced a resolution to

BAS Archives

The second BAS Office at 29 Regent Street.

the property dispute, granting BAS title to the Fort Street property.

But even that was not an unencumbered title, so BAS asked that be corrected. Finally, BAS was granted the desired freehold title to the Fort Street property – Fiat No. 760 of 2003 dated July 7th, 2003.

But by then BAS had begun to out-grow the 12 Fort Street building. In 2003 the office space was increased by adding a container and a gift shop was built thanks to the generosity of Pine Lumber who provided labour and material to be reimbursed from the proceeds of the gift shop. In 2008 the office building was expanded to increase the size of the conference room and in 2009 preparations were underway for the human resources office.

Getting Title to the Fort Street Office

BAS Archives

In 1993 BAS President Therese Rath presents a gift to Lindsay Belisle after he presented the first title to the Fort Street property on behalf of the Ministry of Natural Resources.

Judy Lumb

Pepe Garcia receives the second title from Hon. John Briceño, Minister of Natural Resources at the Opening of the 30th Anniversary Celebrations in 1999.

Fort Street Office

(*above*) BAS Office in 2009; (*below*) Additional office space in container (*left*) and expansion for conference room (*right*). (*Photos pp 37 and 38 by Judy Lumb*)

Office Scenes 2009

Dirk Francisco,
Public Participation and Publicity Coordinator

Delia Noble, Gift Shop Manager and
Programme Support Staff

Dominique Lizama, Protected Areas Manager
(Victoria Peak Natural Monument, Cockscomb Basin and
Crooked Tree Wildlife Sanctuaries)

Anna Dominguez Hoare, Executive Director in 2009

Silver Anniversary Celebration—February 1994

BAS Archives

The Dance, "When", performed by the Belize Dance Company at the 25th Anniversary Opening Ceremony

BAS Archives

Founder Dora Weyer receives an award from Belize's First Lady Mrs. Kathy Esquivel (*right*) with BAS President Therese Rath (centre).

25TH ANNIVERSARY CELEBRATION

The 25th Anniversary Opening and Award ceremony was held at the Bliss Institute on February 5, 1994. The Governor General was the guest speaker. Mrs. Esquivel presented certificates of recognition to the 17 BAS founding members and special certificates to BAS presidents (past and present). A dance entitled "When," especially choreographed for the occasion, was performed by the Belize Dance Company.

Guests then moved to the Holy Redeemer Parish Hall to view "Eco-Fest," an environmental exhibit. An essay competition was held with the "Mission Statement" as the topic. Winners were Andrea McSweaney of Belmopan Comprehensive School and Francis Marin of the University College of Belize.

BAS Archives

Judy Lumb

(*top*) Walkathon celebrating BAS 30th Anniversary.
(*bottom*) The crowd assembled to launch the BAS 30th Anniversary Celebrations at Guanacaste National Park. (front row from left) BAS Trustee Barry Bowen; Hon. John Briceño, Minister of Natural Resources; Valerie Woods, Belize Tourist Board Director; BAS member Svea Dietrich-Ward; BAS Executive Director Osmany Salas; Walter Craig, the first BAS Executive Director (*far right*)

30TH ANNIVERSARY CELEBRATION

The BAS 30th anniversary celebration was launched in grand style on February 6, 1999, with a Walkathon. One hundred and twenty-five sponsored individuals walked in three groups from the BAS Office in Belize City to Guanacaste National Park. One group started from the BAS Office, one from Hattieville, and the last from the Belize Zoo. The Opening Ceremonies were held at Guanacaste National Park with Hon. John Briceño, Minister of Natural Resources and the Environment giving the main address.

At the Launching Ceremony two more protected areas were added to those managed by BAS. An agreement was signed for BAS to co-manage the Blue Hole Natural Monument and the Victoria Peak Natural Monument.

Carlos Santos was the Master of Ceremonies, BAS Executive Director Osmany Salas gave the welcome address, Walter (Micky) Craig gave a summary of BAS history, and both BAS President Jose (Pepe) Garcia and Trustee Barry Bowen greeted the assembled group.

Green Art and Flower Show

As part of the BAS 30th Anniversary Celebration, BAS held a Green Art Show along with the Belize City Garden Club's annual Flower Show. Michael Somerville was the Master of Ceremonies for the Opening of the Green Art/Flower Show. Jean Shaw gave the welcome address and John Masson, of the Belize City Garden Club, officially opened the Show. Artists exhibited their paintings and quilts and donated a portion of the sales to BAS. Artist Terryl Godoy demonstrated his painting techniques by painting an arrangement of orchids in the Flower Show. Michael Somerville gave a slide presentation on the work of BAS.

“Walk for a Green Belize”
BAS 40th Anniversary 2009

40th Anniversary Celebration

On Saturday, April 25th, 2009, BAS celebrated Earth Day with “Walk for a Green Belize” from Belize City to Guanacaste National Park. The major sponsors were the Ambassador of Mexico to Belize, Excellency Manuel Lopez Moreno, and the Embassy of Mexico, Atlantic Bank Ltd, and PACT. The event commenced 6:30 am at the Constitution Park in Belize City where the Mayor, Ms. Zenaida Moya, officially welcomed the participants. The highlight of the event was Leela Vernon’s “Toast to Mother Earth” and the Blue Grass band from the U.S. Embassy in Belize.

(*above*) Group from the Embassy of Mexico in Belize participating in the “Walk for a Green Belize”.

(*left*) Forest Department group participating in the “Walk for a Green Belize”.

(*below*) Arvin Coc leads a bird walk in Guanacaste National Park during Earth Day 2009.

Around 500 people participated in the “Walk for a Green Belize” and it is estimated that a total of 800 bags of trash were collected, representing an average of 16 trash bags per mile. Groups involved included: Belize City Council, Costa Rican Embassy, Dangriga Youth Group, Forest Department, Golden Citizens, Guatemalan Embassy, Love FM, Simeon Lopez (Mayor of Belmopan), Mexican Embassy, OAK, PACT, Palotti High School Nature Protectors, Scouts Association, SMART, University of Belize Environmental Club, U.S. Embassy, WWF, and Belize Audubon Members and Staff. Additionally logistical assistance was rendered by the Belize Red Cross, Old Belize, Police Department, Police Recruits, and St. John’s College First Aid Team.

"Shutter Speed: A photography exhibition showcasing Belize Audubon Society at work" Kicks off BAS 40th Anniversary 2009

Judy Lumb

Judy Lumb

Shutter Speed: A photography exhibition showcasing Belize Audubon Society at work

BAS kicked off a year-long 40th anniversary celebration with the opening of "Shutter Speed: A photography exhibition showcasing Belize Audubon Society at work" at the Mexican Embassy in Belize City on February 5, 2009. The Exhibition was opened by BAS Vice President Michael Heusner; Licenciado Luis Manuel Lopez Moreno, Ambassador to Belize from Mexico; Manuel Rubido, Ambassador to Belize from Cuba, all three cutting the ribbon (*photos above*). Attendees were entertained by the music of Mark Phillips and his jazz trio. Snacks and drinks were served.

The 104 beautiful photographs are the creative work of Dirk Francisco, BAS Publicity Coordinator (*photo left*). For the rest of 2009 the exhibit travelled throughout Belize to Belmopan, Dangriga, Benque del Carmen, Punta Gorda, Orange Walk, Corozal, and even Chetumal, Mexico. Photos from this exhibit are shown throughout this book.

"Shutter Speed: A photography exhibition showcasing Belize Audubon Society at work" Kicks off BAS 40th Anniversary 2009

(*left*) Shutter Speed banner; Dudley Heredia, BAS Financial/Office Manager, and BAS Director Lascelle Tillett (*above*) and Sarita Tillett (*below*) view the exhibit; Mark Phillips Jazz Trio (*below left*) entertains at the Opening Ceremony (*photos by Judy Lumb*)

"Shutter Speed: A photography exhibition showcasing Belize Audubon Society at work" Kicks off BAS 40th Anniversary 2009 (cont'd)

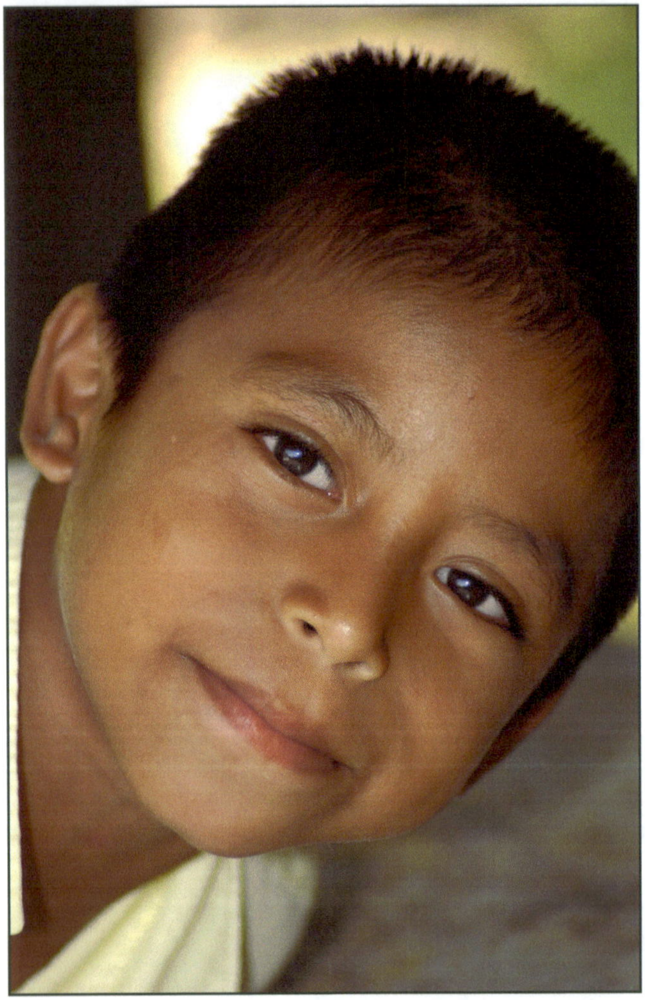

Blue Morpho (*Morpho peleides*)

"Shutter Speed: A photography exhibition showcasing Belize Audubon Society at work" Kicks off BAS 40th Anniversary 2009 (cont'd)

Andrea Coc

Phalaenopsis amabilis

"Shutter Speed: A photography exhibition showcasing Belize Audubon Society at work" Kicks off BAS 40th Anniversary 2009 (cont'd)

Perry Diaz at the 2006 Summer Camp at the Ben's Bluff Waterfall in Cockscomb Basin Wildlife Sanctuary.

Tiger Fern Waterfall

"Shutter Speed: A photography exhibition showcasing Belize Audubon Society at work" Kicks off BAS 40th Anniversary 2009 (cont'd)

Parrot Heliconia (*Heliconia psittacorum*)

Orchid --- not yet identified

"Dancing and Dining with Nature"
BAS 40th Anniversary 2009

An Eco-Fashion Design and Modelling Presentation was presented by Emil Alvarado featuring clothing made from recycled materials such as garbage. Some models were BAS staff Olivia Carballo-Avilez (wedding gown), Dominique Lizama (bubble wrap dress) and Khaidia Mossiah (black garbage bag dress).

40th Anniversary Gala

Celebrating the 40th anniversary, BAS members and friends enjoyed the most memorable gala of the year under the theme of "Dancing & Dining with Nature, held at the Belize Biltmore Plaza Hotel on June 20, 2009. A three-course tropical meal was served with wine. Music was provided by the Youth Connection Band, as well as entertainment by Tremett Periott, Pablo Collado, Sarita Moralez, and Ameera Marin. BAS thanks our main sponsors for the Gala, Atlantic Bank Ltd, Embassy of Mexico in Belize, Belize Tourism Board, Belize Telemedia Ltd, Love FM, and PACT.

A dance was presented by Alma Caribeña courtesy of the Mexican Embassy.

Enjoying the 40th Anniversary Gala are BAS President Earl Green, his wife Ida, the Governor General Sir Dr. Colville Young, and his son BAS Director Carlton Young.

Summary—Formation and Growth of the Belize Audubon Society

On February 6th, 1969, the forty people gathered at the Premier's Lodge decided to form the Belize Audubon Society "to study wildlife and its conservation". Belize Audubon Society was established as a Chapter under Florida Audubon's International Committee.

Others joined the effort so that by the next meeting on March 7th, 1969, there were 55 charter members. Officers were elected and six committees established: Field Trips, Programme, Conservation, Membership, Youth Education, and Sanctuary.

The Society met every month and members took charge of the programme bringing their expertise to the membership. In order that members could learn about Belize's wildlife, field trips were organized. Members enjoyed field trips to locations that later became protected areas, such as Crooked Tree and the inland Blue Hole. Trips were made to Half Moon Caye to try to rid the island of rats that were a threat to the Red-footed Booby chicks.

It is often stated that BAS started as a bunch of bird-watchers. While the founders were involved in bird-watching, the six committees established from the very beginning indicate that Belize Audubon Society has always been committed to advocacy (Conservation), environmental education (Youth Education), and protected areas (Sanctuary).

The growth of the Society can be measured in several ways. The membership has grown from the original 55 founding members to over 1200. The professional Staff has grown from none for the first fifteen years to over 40 at the 40th anniversary. The budget has grown from a few hundred dollars to over two million.

For the first fifteen years BAS functioned as the only environmental organization in Belize solely through the efforts of devoted, hard-working volunteers. By the dawn of 1985 a new government had been elected and the Society was joined by the Belize Centre for Environmental Studies, the Belize Zoo, and, a couple of years later, Programme for Belize. BAS had been called upon to manage several protected areas and funds were becoming available to open an office and develop a professional Staff.

BAS developed a vision for the 1990s which included a focus on protected areas management in the first strategic plan, but before that planning cycle was completed, environmental education and advocacy had been added.

By the second strategic plan, developed in 2001, BAS had developed enough of a reputation that the public expected much, from wildlife management to the control of development. There was a concern that BAS was trying to do too many different things, so the decision was made to focus the work on protected areas management with environmental education and advocacy supporting the work at those protected areas.

By the new millennium, it had become clear that there is no true protection without the consent and participation of the local people, those with a direct stake in the management of protected areas. BAS was focusing on the buffer communities surrounding the BAS-managed protected areas and the stakeholders of the marine protected areas. So the third strategic plan for 2008–2012 includes community work, as well as acknowledgement of ongoing research and other conservation work supporting the preservation of biodiversity.

With each new milestone, the Society celebrates: 25 years, then 30 years and now 40 years of conservation in Belize!

Richard Pavek

BAS Member Richard Pavek photographed this Jabiru "love dance" in January of 2007. The female was alone on the nest, and then the male Jabiru landed, settling his feet on top of the female's shoulders. The male then did a little dance on her shoulders, lightly hopping from one foot to the other, for about a half a minute.

Belize Audubon Society

Celebrating 40 Years

Chapter Two

The Advocacy Programme at the Belize Audubon Society

National Parks and Wildlife Protection Legislation

Lydia Waight

Even before BAS was formed, the founders were working for conservation in Belize. In 1966 a National Parks Commission was constituted to travel around Belize and identify areas to be recommended for protection as national parks. Among the members were Henry Fairweather, Albert S. Grant, Louis Lindo and James Waight. Ronald Clark was secretary to the Commission. These five men later became BAS officers and Board members. The Commission's report was completed in 1968. After its formation the next year, the BAS lobbied on behalf of national parks.

On September 21, 1981, Belize became an independent country. Two of the first acts of the new country were the Wildlife Protection Act 1981 (No. 4 of 1981) and the National Parks System Act 1981 (No. 5 of 1981), which were signed on November 13 and published in the *Honduras Gazette* of November 28. Both were major achievements for BAS because members had worked long and hard on these landmark pieces of legislation.

The Wildlife Protection Act came into force on January 2, 1982. The Act provides for the conservation, restoration, and development of wildlife, for the regulation of its use, and for all other matters connected thereto. Under the Act, the term "wildlife" means all undomesticated mammals, birds and reptiles, and all parts, eggs, and nests of any of these forms. A list of wildlife species, the handling of which is prohibited, is included with the Act. These animals are considered rare and/or endangered, many of them listed as needing protection in the Convention on International Trade of Endangered Species of Flora and Fauna, known as the CITES Convention, of which Belize is a signatory.

Among the animals protected by the Act are our monkeys, the spider monkey and howler monkey (locally known as "baboon") and our five wild cats. Also included are the Central American or Baird's tapir (the "mountain cow," our national animal), the manatees, our two crocodiles, our dolphins, and any whales that venture into our waters. Our birds, with few exceptions, are also protected, and may not be hunted or captured for sale as pets.

There are provisions, of course, for any person to take measures to defend himself or other persons from the attack of any animal, and for owners of crops or domestic animals to kill or capture on their property any wildlife threatening or causing material damage to such crops or domestic animals.

To hunt our common game animals, a traditional food source—deer, peccary, gibnut, armadillo, and the game birds, curassow, cocrico, quam, and quail—requires a valid hunting license issued by the Game Warden.

The Wildlife Protection Act also prohibits the hunting of any immature wildlife, or any female accompanied by its young. This means it is illegal to capture young animals for the pet trade. A moratorium on commercial dealing, that is, on the sale or dealing for profit in any wildlife of any species, or part or product thereof, went into effect from the day on which the Act came into force.

The Wildlife Protection Act was a big step toward preventing the destruction of our natural heritage. If we are to keep our wildlife for future generations to use and enjoy, that wildlife needs to be protected, and its use controlled. Many countries have realized too late that their wildlife has been destroyed by over-hunting and the clearing of forests. We must not let this happen here!

Wildlife Protection

Jabirus

The Jabiru was BAS' first conservation project. One of the first acts of the BAS Board of Directors was to write a letter to the Acting Chief Forest Officer and the Chief Game Warden. The *BAS Bulletin* included the following report:

The Board of Directors of the Belize Audubon Society, the aim of which is to conserve the wildlife of our country, is seeking your help for the protection of the Jabiru Stork. The reason for this request is that the Jabiru is very uncommon; it is rare in Central and South America and may soon become extinct if left unprotected. ...

The Belize Audubon Society respectfully requests that the jabiru, Jabiru mycteria, *should be put on the list of protected birds making it an offence against the law to kill or capture the bird or to rob its nest of the eggs. If protected here, this bird could be a big tourist attraction. (Letter to the Acting Chief Forest Officer and Chief Game Warden, June 12, 1969)*

In 1973 the Jabiru was added to the list of protected animals (*Gazette* 7 July S. I. No. 38/1973).

Manatees

BAS got a report that fishermen from Ambergris Caye and other places were killing manatees. So, BAS wrote to the Chief Game Warden to ask him to remind the "police officers of San Pedro, Caye Caulker, and Sarteneja, who are deemed game rangers under the Wildlife Protection Ordinance, that the manatee is a protected animal and it is prohibited by law to kill these animals."

Export of Hides

BAS had representation on the committees controlling the export of hides and other wildlife products. Hides of cats were the first to be prohibited from exportation.

Parrots

BAS kept alert to any threat to Belizean wildlife. When parrots were offered for sale on a radio programme in 1976, BAS alerted the Chief Forest Officer and suggested to the Chief Broadcasting Officer that

Manatee photos by Fred Sharrock

advertising of this nature should not be broadcast over Radio Belize. The Chief Forest Officer followed up with a request to the Chief Broadcasting Officer that no announcements be made for sale of animals except domestic ones.

The Society wrote not only to complain, but also to commend officials for their positive actions. In 1977 the Society received a report that the export of a large number of Yellow-headed and other parrots had been stopped and wrote to commend the officer for his prompt action to prevent these parrots leaving the country.

In May of 1987 BAS was informed that 5,070 live birds in 55 wooden boxes had left Belize on their way to Amsterdam, Holland, via Costa Rica. After investigations and discussions with Government authorities, the Society was able to establish that the birds had not been taken from Belize. The documents were shown to be false. The Society "viewed this fabrication was a malicious and deceitful act that damages the good name of the Government and people of Belize."

Hunting

Concern was raised in 1976 over the large numbers of hunters shooting ducks at Big Falls by BAS Vice President W. Ford Young. BAS was concerned about resident ducks and geese that were becoming quite rare. The Belize Audubon Society feels that the Black-bellied Tree-Duck (*Dendrocygna autumnalis*) and the Muscovy (*Cairina moschata*) should not be shot.

Five years later BAS was still concerned with hunters. The following are excerpts from a radio spot that was broadcast on July 14, 1981.

> *A favourite stop on the Belize Audubon birding trips to the Burrell Boom/Bermudian Landing area is the bridge over Mussel Creek. ... Just as it begins to grow light, flights of herons and egrets start moving to their favourite feeding grounds. The call of the mottled wood owl can be heard. ...*
>
> *Chattering parrots fly overhead and the harsh croak of the "barking gaulin" (the bare-throated tiger-heron) reveals its presence. The jacana (our "georgie bull"), the gray-necked wood-rail (we call it "top-na-chick"), snail kites, kingfishers, the limpkin (which we call "clucking hen"), ...*
>
> *One recent Sunday afternoon, Audubon Society members, returning from a field trip, met up with a party of shooters who had been killing "shegs" (our*

local name for cormorants). It does not take a great marksman to hit these unwary birds! Like other fish-eating birds, shegs are not very palatable, so the dead birds are just left lying around.

Not only cormorants are targets for these shooters. A reliable report tells of a farmer fishing out dozens of dead wood storks from Mussel Creek, left there by Sunday shooters. To make use of them, he fed them to his pigs! The wood stork (our "john crow curlew" or "culu") is a protected bird and its nesting rookeries have been declared bird sanctuaries.

Should this wanton and indiscriminate shooting be allowed to continue, the birds will soon be gone. Those that are not killed, or wounded and left to die, will be scared away by the noise of the guns, and a favourite spot for bird watching will be lost. Not only the Belize Audubon Society, but other citizens as well, would never like this to happen. In an attempt to prevent this disaster, an earnest appeal is made to all hunters. Please stop shooting at Sebastian Bridge, so that the bird life of that area may be left undisturbed.

BAS Archivs

Scarlet macaw (*Ara macao*)

Scarlet Macaws

Reports reached the office of the Belize Audubon Society on February 19, 1997 that unknown persons from the village of Red Bank in the Stann Creek District were involved in the shooting of scores of Scarlet Macaws (*Ara macao*).

A rapid response from the environmental NGOs and the Government of Belize turned this tragic discovery into a tourism opportunity. The killing of scarlet macaws stopped immediately, as soon as the people of the village realized that it was illegal. Through a grant from the United Nations Development Programme, Global Environmental Facilities, Programme for Belize helped the people of Red Bank Village to develop a guest house and trails. Tour guides were trained and tourists now go there to see Scarlet Macaws.

Confiscation of illegally Captured Birds

Fifteen wild birds were confiscated from a pet store in a private residence in Belmopan by officials from the Police and Forest Departments. BAS member and expert birder Philip Balderamos was called to identify the confiscated birds.

PROTECTING MARINE RESOURCES

Sea Turtles

Beginning in 1990 an extensive sea turtle programme has been carried out in Belize, spearheaded by BAS member Greg Smith. Volunteers patrolled the beaches where sea turtles nest in north Ambergris Caye and, later, Manatee Bar in the Gales Point area. At Manatee Bar he found one of the largest nesting sites for hawksbill turtles in the Caribbean region.

This programme was supported by the U.S. Fish and Wildlife Service through the Reef Preservation Committee. The Gales Point field site was established in cooperation with G. Winston Miller of the Fisheries Department under a grant from the World Wildlife Fund. The "Belize Sea Turtle Recovery Plan" was completed and published in 1993 with the assistance of Karen Eckert of Wider Caribbean Sea Turtle Conservation Network (WIDECAST). Lobbying by BAS led to passage of laws protecting sea turtles. Sea turtles are now completely protected, so that fishing, selling, buying, and importing any part of sea turtles are prohibited by Belize law.

Carlos Ayala

Green Sea Turtle (*Chelonia mydas*)

Leasing the Reef

A private company made a proposal to lease a portion of the Belize Barrier Reef for tourism activities. BAS and many others in both conservation and tourism expressed their concern about this proposal. The Government of Belize noted this overwhelming stakeholder support for government management of marine resources rather than private enterprise.

Banning Shark Feeding in Blue Hole

For some time BAS has been advocating for a ban on shark feeding at Blue Hole Natural Monument. The practice is dangerous and is changing the feeding habits of the sharks. This campaign has involved lobbying the Belize Tourist Board and the Forest Department to support the Society's position. Posters have been created to target tour operators, tour guides, and tourists. The goal of this campaign is legislation banning this practice with enforcement mechanisms.

Whaling Commission

BAS was asked by the Government of Belize to select a representative to the Whaling Commission. With support from the Tinker Center for Coastal Studies in Latin America, BAS member David Craig attended the 35th Annual Meeting of the International Whaling Commission held in Brighton, England, in July of 1983.

At the 57th sitting of the Whaling Commission in 2005, Japan put forward three amendments that would establish voting by secret ballots, re-opening of commercial whaling in Antarctica, and increase quotas on the basis of economic, social and cultural needs. BAS lobbied the Government of Belize (GOB) to vote against these three amendments. While some Caribbean representatives voted with Japan, the GOB announced that they would not support the Japanese initiatives. The amendments were defeated by a small margin.

MV Transfer aground at Lighthouse Reef (*photos by Kirah Forman*)

Sun Jam runs aground at Half Moon Caye

Threats to Lighthouse Reef

Roberto Pott

At about 11:30 pm on the 6th of January, 2005, a cargo ship ran aground on the reef, four miles southeast of Sandbore Caye in the northern area of Lighthouse Reef. The 225-foot ship is registered in Antigua and Barbuda as *Motor Vessel Transfer*. The *Transfer* had arrived at Big Creek at about 8:15 am on the 5th of January, unloaded cargo that included some hardware and supplies for shrimp farms. Her next port of call was to have been Mobile, Alabama, but she ran aground while reportedly running under automatic pilot. None of the crew of eleven was injured.

BAS liaised with the team including members from Belize Ports Authority, Coastal Zone Management, Department of Environment, Esso Standard Oil and a Surveyor to monitor the situation, headed by Major G. H. Swaso. The fuel included over 11,000 gallons of Diesel fuel, which was removed.

On the 25th of the same month, a passenger cruise vessel, *Le Levant*, attempted to tie up to a mooring, but the mooring was not strong enough for the vessel and the mooring was dislodged, damaging nearby coral.

Then a 38-foot sailing catamaran hull ran aground within the Half Moon Caye Natural Monument, a protected area co-managed by BAS. The catamaran, *Sun Jam,* is registered in California and was being captained by David Luckner. BAS, along with a team comprised of the Department of the Environment, Belize Port Authority, Coast Guard and the Customs and Immigration Departments assessed the damage.

Pesticides

In 1974 Gramaxone (paraquat) poisoning was reported in the cane workers in the north of Belize. The Society was concerned because there was no legislation yet in Belize controlling the sale, import or use of pesticides. BAS wrote for information and passed it on to Government.

BAS was again concerned with potential pesticide use when a large-scale cotton cultivation project was proposed for Belize. Cotton was being raised in Nicaragua and workers had died from pesticide poisoning. Again, BAS passed information to the Government. This project was never approved.

The Pesticide Control Act was gazetted on 28 December 1985. Later on there were Statutory Instruments setting out regulations for the use of controlled pesticides.

National Conservation Strategy

After independence, Belize had international help in developing plans for conservation, including a mission consisting of three members of the IUCN Commission for Environmental Planning – Pierre Hunkeler, Cyrille de Klemm, and Merilio Morell; as well as Archie Carr III of the New York Zoological Society, Bruce Rich of the Natural Resources Defence Council, and R. Dubois of the USAID. Their report, "Conservation for Sustainable Development in Belize: Towards a National Conservation Strategy," was presented to Government in May of 1983. The next year Mark Halle of IUCN and James Barborak and Craig MacFarland of *Centro Agronomica Tropical de Investigacion y Enseñana* (CATIE) came to develop project proposals to provide technical assistance, training and funding. CATIE intern David Lovell prepared draft management plans for Cockscomb Basin Wildlife Sanctuary, Crooked Tree Wildlife Sanctuary, and Half Moon Caye Natural Monument.

National Protected Areas Policy and System Plan Project

Twenty years later BAS provided financial and technical assistance for a major policy and planning effort, which resulted in the development of the National Protected Areas Policy and System Plan Project (Plan). The Plan was approved by Cabinet in December, 2005. The purpose of the Plan was the development of a comprehensive national protected areas policy addressing issues such as:

- reservation, de-reservation, and co-management;
- financing and management frameworks; and
- the role of protected areas in national development.

Game Ranch Project

In the midst of the elections of 1984, a project was proposed for Belize that caused considerable controversy and conflict within BAS. Kent B. Crane proposed to develop 20,000 acres of marginal land near Crooked Tree as a game ranch where foreign hunters would come to shoot exotic animals imported from Africa and other places, along with native wildlife. He also proposed a captive breeding programme for endangered species using embryonic transfer, stating that it was in collaboration with the Smithsonian Institute. An Environmental Impact Assessment was being done. Crane came to a BAS Board meeting and brought Jan Oelofse, who had a similar game ranch in Namibia, Africa.

BAS raised concerns about the introduction of exotic animals to Belize and sought advice from friends in the U.S., but the BAS Board was divided on the issue. Some were in favour because of the breeding programme for endangered species. Others urged caution. The GOB was said to be agreeable with the approval of the Chief Veterinary

Officer and BAS sent a letter to the Minister of Natural Resources expressing cautious support of the two projects.

But then BAS saw the Crane Game Ranch proposal and received replies from experts at the New York Zoological Society, who expressed grave concerns about Crane's project. They also learned that Crane was not involved with the Smithsonian.

BAS had just received the first grant from the World Wildlife Fund to support the new Cockscomb Basin Wildlife Sanctuary. It was suggested that if BAS supported the game ranch, they might lose the WWF funding. The BAS Board then reversed its position and so informed the Minister of Natural Resources.

Coca Cola

On October 3, 1985, a news release from Coca-Cola Foods of Houston, Texas, announced their plans to invest in Belize, approximately 700,000 acres of the property owned by the Belize Estate and Produce Co. Ltd., in northwestern Belize.

This Coca-Cola announcement prompted a strong response among environmentalists and international funding agencies that had been investing in various projects to insure that Belize's development could proceed in a sustainable, environmentally-friendly manner.

Friends of the Earth reported in a BBC programme that Coca-Cola Foods claimed to be working closely with BAS and implied that BAS would be doing the Environmental Impact Assessment (EIA) for the project. BAS released a statement on January 8, 1987, indicating that, except for an initial meeting, CCF had not worked closely with BAS and that BAS had neither the human nor financial resources to do an EIA. CCF responded with an apology.

After a meeting with Coca-Cola Foods, the BAS Board prepared the following policy statement.

Because of its relatively untouched natural environment, the Government and people of Belize have the unique opportunity to design and implement a comprehensive strategy for sustainable development that integrates concepts of environmental conservation in the planning process. The Belize Audubon Society is cognizant of the need for economic development of the country so as to ensure the wellbeing of the people. The urgency of the need for development, however, may threaten to encourage measures that produce solely short-term financial gains rather than policies designed to achieve long-term, environmentally sustainable development. Thus, to ensure that the development of Belize results in long lasting, positive change, the Belize Audubon Society urges those engaged in activities relating to the further development of the country to keep the value of natural areas in the forefront of their thinking and fully integrate environmental protection measures into development policies.

At a meeting held on 19 March 1987, between personnel of Coca-Cola Foods and the Belize Audubon Society, Coca-Cola Foods personnel explained as best they could their present position in the development of citrus groves in the northwestern sector of the country of Belize.

Coca-Cola Foods is prepared to fund the development of an environmental impact assessment of the area to be cultivated and has agreed to have the involvement of the Belize Audubon Society at all phases of development of the said impact statement.

Consideration will also be given to the possibility of setting aside a parcel of land which would be left undisturbed so as to ensure the preservation of the flora and fauna native to the area.

The need for continued dialogue between Coca-Cola Foods and the Belize Audubon Society was recognized and assurances were given the Society that efforts would be made to see that this is done.

The Belize Audubon Society appreciates the interest of Coca-Cola Foods in addressing the environmental issues associated with the citrus project and looks forward to a fruitful and meaningful working relationship. (BAS Press Release, 10 April 1987)

The issue was covered on "60 Minutes," in the United States, and there was extensive coverage in Europe. The media exposure brought considerable international attention to environmental issues in Belize and to the Belize Audubon Society.

Fortunately, the conservation movement in Belize is now a powerful force, thanks to the naturalists' organisation, the Belize Audubon Society, and its extraordinary influence on a sympathetic government. The separate initiatives of the Belize Audubon and other organisations and the government have consolidated this achievement.

In 1981, the government passed the National Parks System Act. Under the act, the Audubon Society has a mandate from the government to help it to develop and run the parks. Since 1981, five parks have been established, and more are planned. . . . Dean Lindo, the agriculture and conservation minister, says, "We're going at conservation hell's bells. This is the uniqueness of Belize that we can make decisions now to preserve and protect what we have." (New Scientist, *29 October 1988)*

In the end Coca Cola pulled out altogether. The Rio Bravo Conservation and Management Area was established with land donated by Coca Cola and adjacent land purchased by the Massachusetts Audubon Society under the management of Programme for Belize.

Logging in Toledo District

The first logging controversy to arise in regard to the Columbia Forest Reserve in the Toledo District involved a Mexican company in 1992. The environmental NGOs joined forces in a successful lobbying effort against this plan.

Then in 1994 a logging concession for 24,000 acres of the Toledo District was given to Atlantic Industries Corporation, a Malaysian firm. There was such an outcry raised by the Belizean environmental organizations with support from international organizations that the project was stopped, but not for long. In 1996 the project was back again, this time with a modern forest management plan revolving around selective logging of small blocks of forest and then allowing that block to recover for several years. There was to be no cutting within 60 feet of rivers or streams. The permit covered 18 percent of the Columbia Forest Reserve and areas along the boundary of the Sarstoon-Temash National Park.

Again there was an outcry from environmental organizations and also from the local indigenous people, the Mopan and Kekchi Maya, who organized massive protests. BAS conducted a fact-finding mission to the area in June of 1996 and found that the loggers were violating the terms of the approved plan by taking too many trees from too large an area and by cutting along riverbanks.

In September of 1998 the GOB declared a halt to the logging until adequate consultation with the local population had occurred. A committee chaired by BAS President Pepe Garcia was appointed to review the Malaysian logging licenses. The report submitted to the GOB in March, 1999, made the following recommendations.

1) The license should be cancelled, redefined to address existing concerns, and renegotiated.

Judy Lumb

Judy Lumb

Legal, sustainable logging in the Toledo District carried out by local indigenous people.

2) If negotiations cannot be concluded with the present operator, local industry should be given priority in obtaining the license over other foreign or new investment.
3) The license area should be redefined including:
    a) A buffer zone established around each community for traditional uses with community involvement in the process.
    b) Protected forest and other areas of national heritage must be removed from areas open to harvesting.
    c) The remaining productive forest must be placed on an area control basis with not only technical sustainability, but also economical viability duly considered.
4) The operator of the license should be required, to truly effect positive economic impact on the region as was promised.
5) The license was issued without utilization of expertise in the department, projects and other relevant government offices. Appropriate disciplinary measures must be taken and measures taken in the department to avoid such abuse in the future.
6) The supervision by the Forest Department seems to have been very inadequate. The Forest Department presence in Toledo must be upgraded to a full territorial division with the necessary allocation of resources.
7) Incorporate larger annual compartments to increase economic viability of the licensee in carrying out the demanding conditions of a sustainable yield license.
8) Prior to the issuance of a license, the impact on the surrounding communities and the environment, should be properly addressed and incorporated in a management plan for the license area.

National Environmental Appraisal Committee

Nimmi Herrera and Tanya Williams-Thompson

The Environmental Protection Act (EPA) No. 22 of 1992 created the Department of the Environment (DOE) as the agency charged with ensuring the protection and sustainable use of Belize's natural resources. The EPA made provision for the Environmental Impact Assessment Regulations of 1995, which included the creation of the National Environmental Appraisal Committee (NEAC), the advisory body charged with reviewing all environmental impact assessments (EIAs).

Membership on the committee included:

- Chief Environmental Officer or his nominee
- Commissioner of Lands or his nominee
- Housing and Planning Officer or his nominee
- Chief Forest Officer or his nominee
- Fisheries Administrator or his nominee
- Chief Hydrologist or his nominee
- Archaeological Commissioner or his nominee
- Director of Geology and Petroleum or his nominee
- Chief Agricultural Officer or his nominee, and
- Two non-governmental representatives (NGO) appointed by the Minister on the recommendation of the Department.

BAS served as one of the NGO representatives, first as a representative of the Belize Association of Conservation NGOs (BACONGO), and later as a representative of the Association of National Development Agencies (ANDA). In 2007 BAS was appointed for a two-year term as an independent NGO, which ended in 2009.

For 13 years BAS participated actively in reviewing the EIAs for developmental projects considered by the NEAC, usually 10–15 projects per year. The government representatives were capable technical personnel who provided sound advice in each speciality area. However, they also represented the development policies of the government, which sometimes was incongruent with the conservation policies of the two NGOs. Because there were only two NGOs, they were outnumbered by the government representatives. Another difficulty was that the regulations limited the review to environmental impacts and did not include socioeconomic impacts. BAS lobbied to change the regulations to include additional NGO representatives on NEAC and to consider socioeconomic impacts of the projects being reviewed.

Other Committee Membership

Over its 40-year history BAS has served on many other national and regional committees, including, but not limited to:

- Association of National Development Agencies
- Belize Barrier Reef Committee
- Coastal Zone Advisory Council
- Energy Policy Steering Committee
- Fisheries Advisory Board
- Marine Protected Areas Working Group
- National Biodiversity Policy Committee
- National Coordinating Body for UN Convention to Combat Desertification
- National Energy Advisory Council
- National Land Advisory Committee
- National Pro Tempore Water Commission
- Protected Areas Conservation Trust Advisory Council
- Senate Advisory Committee

Lamanai Room Declaration

The year 1997 was known internationally as "The Year of the Reef," but in Belize it brought a number of threats to our beautiful, fragile reefs and natural resources. Two diving boats dragged their anchors over the reef at Lighthouse Reef in July, a captive dolphin theme park was approved, exceptions to the lobster fishing laws were approved to allow for lobster ranching, and there were rumours of live rock exportation. These threats to Belize's most precious and unique natural resource prompted BAS to hold an unprecedented joint Board of Directors meeting with the Belize Tourism Industry Association (BTIA). They invited other environmental organizations as well to the Lamanai Room at the Radisson Fort George Hotel. The resulting Lamanai Room Declaration, was signed by 26 organizations on August 7, 1997.

Whereas, in a joint meeting, the BTIA and the BAS, with the support of numerous non-governmental organizations, after due deliberations, have concluded in preliminary analysis that actions taken by the Government of Belize ... are indicating disregard for principles of sustainable development, community consultation, conservation and environment, and are indicating a possible violation of Belizean environmental legislation ...

Be it resolved that we the undersigned shall:

1) *jointly retain legal counsel to represent the multi-sectoral interest, and the interest of all Belizeans;*
2) *support the position of the fishing cooperatives of Belize on lobster ranching and other reef-related issues;*
3) *advise the international environmental community in respect of the above;*
4) *not discount the possibility of dialogue with international funding agencies.*

A second meeting between BAS) and BTIA was held on October 21, 1997, at the Radisson Fort George Hotel in Belize City and the following update was issued:

1) **Live Rock Exportation**: It was reported that the Government of Belize (GOB) has not received any application nor granted any concession for the exportation of live rock.

2) **Reef Damage:** The Ministry of Tourism and the Environment has collected a fine of BZ$75,000 from each of the two vessels that caused the recent damage to the reef. The Statutory Instrument relating to environment protection is currently being amended to increase the fines for reckless damage to the reef.

3) **Logging in the Toledo District:** It was agreed that a closed meeting will be held between the BTIA and BAS to study all the parameters surrounding the issue and that all stakeholders will be invited.

4) **Lobster Ranching:** Two licenses were granted for lobster ranching, one to A&J Lobster near Cosiga Creek (James Wang Company), and a second to Dangriga Taiwan Initiative Ltd. near Bread and Butter Caye in the Stann Creek District. These projects were eventually cancelled.

5) **Dolphin Theme Park at Cangrejo Caye:** BTIA/BAS officials met with the Minister of Tourism and the Environment, the Permanent Secretary, the Chief Environmental Officer, and the Director of Tourism. The Chief Environmental Officer stated that the Cangrejo Caye investors were given the terms of reference for a required Environmental Impact Assessment Study. The DOE agreed to consult with BTIA/BAS on these matters. This project was eventually cancelled.

Lamanai Room Declaration II (*from left*) BAS Advocacy Manager Tanya Williams-Thompson, Belize Tourism Industry Association Executive Director Andrew Godoy, BAS Executive Director Anna D. Hoare, BAS Vice President Mike Heusner, Healthy Reefs Executive Director Melanie McField, and Programme for Belize Executive Director Edilberto Romero

Lamanai Room Declaration II

Tanya Williams-Thompson

In 2007 BAS confirmed that the Government of Belize planned to de-reserve portions of Bacalar Chico National Park and fourteen other protected areas. These issues were used as a springboard for the Lamanai Room Declaration II.

Ten years after the first Lamanai Declaration in 1997, it seemed that Belize was faced with even greater issues that threaten the health of our environment, our culture and heritage, and our livelihoods. In this historical context BAS, BTIA, APAMO, and SPEAR undertook a process that engaged all interested individuals in drafting the Lamanai Room Declaration II. The public was invited to the Endorsement Day presentations on the issues and more than 250 endorsed this declaration:

RECALLING the Lamanai Declaration of August 13, 1997, and its importance to the development of a harmonious relationship between environmental protection and meaningful sustainable development; ...

*THEREFORE, [*BAS, BTIA, APAMO, SPEAR*], and the undersigned organizations and individuals;*

CALL UPON all Belizeans to become actively involved in efforts to protect and preserve the environment and natural resources of Belize; and

STRONGLY URGE the Government of Belize to immediately:

Develop and legislate a systematic and transparent, public process of declaration, alteration, and de-reservation of protected areas;

Institute a formal moratorium on the designation and de-reservation of protected areas until the above action is completed;

Provide the necessary legal, human and financial resources and support to stop illegal activities and incursions in Belize's protected areas;

Protect Belize's territorial integrity by actively and decisively addressing the agricultural incursions and other encroachments on Belize's territory by Guatemalan citizens who are depleting and degrading our natural resources at unprecedented levels;

Revise the Environmental Impact Assessment Regulations through broad stakeholder consultation with the aim of making the process more effective, participatory and representative of all stakeholders;

Finalize and legally adopt the Belize Coastal Zone Management Plan which includes analysis of cumulative impacts and provides a blueprint for development in the coastal and marine areas;

Increase financial support for the enforcement of environmental regulations including Environmental Compliance Plans, dredging and mangrove clearance permits, among others;

Implement the solid waste management plan financed by the Environmental Tax;

Develop and implement a sustainable tourism strategy, with broad stakeholder participation, which levels the playing field for national and international investors and provides incentives that reward green investment, and,

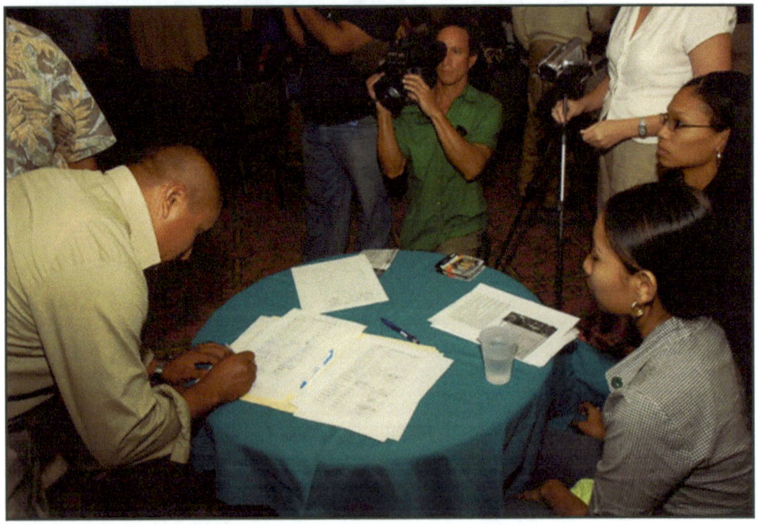

(*top*) Capacity crowd at the Lamanai Room Declaration II
(*bottom*) Signing the Declaration

Develop a long-term national development plan, with the full participation of environmental stakeholders, ensuring that this plan provides for sustainable economic development of Belize while safeguarding our natural resources.

Lamanai Room
Radisson Fort George Hotel
November 8, 2007

HIVOS Grant

The goal of the BAS Advocacy Programme is to take action to ensure that policies set forth by the Government of Belize, especially with regard to land and water, are environmentally sound and take the best interest of the Belizean people into consideration. This programme was funded by a Dutch NGO, the Humanitarian Institute for Cooperation with Developing Countries (HIVOS), beginning in September, 1998, which allowed BAS to hire an Advocacy Coordinator. Major objectives of the programme were:

- To establish an expanded grass roots base for BAS,
- To advocate for policies and enforceable regulations that take into account sound environmental behaviours amongst specific targeted groups,
- To establish an information and research system to complement and add to the Conservation and Environmental Data System,
- To take a leading role in steering the Belize Alliance of Conservation Non-governmental Organizations (BACONGO) and participate actively in the Association of National Development Agencies (ANDA), and
- To demonstrate successful approaches to sound environmental behaviour.

HIVOS support for the BAS Advocacy Programme continued from 1998 through 2007, supporting much of the work discussed in this chapter. BAS is very grateful for this support.

An Environmental Agenda for the 21st Century

Valdemar Andrade

Under its Advocacy initiative, BAS launched a document entitled "An Environmental Agenda for the 21st Century." This document outlined BAS' position on the main issues affecting our natural environment, some of which were:

- People and the Environment
- Tourism
- Environmental Education
- Fisheries and the Coastal Zone
- Waste Management
- Forestry and Protected Areas
- Water Resources
- Land Ownership and Development
- Agriculture

The document gave the current status, pointed out the challenges/concerns and offers proposals for action. This document involved several months of preparation and consultation with Belize professionals in their relevant fields. The Agenda was officially launched on 23 June 1998 to the media, politicians, NGO's, development organizations, government departments, and the general public.

As a result of this effort and for the first time in Belize electoral history, both the People's United Party and the United Democratic Party Manifestos had whole sections relating to the environment, ecotourism, and management of our natural resources.

Environmental Agenda 2002 and Beyond

In April of 2002 BAS launched *Environmental Agenda 2002 and Beyond: An Update of an Environmental Agenda for the 21st Century*. Again a team of experts wrote initial anonymous reports which were reviewed by the BAS Staff and Board of Directors. In addition to the subjects covered in the first *Environmental Agenda*, Energy Generation and Aquaculture were added; Fisheries was covered separately from the Coastal Zone; and Forestry was covered separately from Protected Areas. In addition to the printing of the document, sections were published in the *BAS Newsletter* as appropriate to the theme.

An Environmental Agenda for Belize 2008-2013

Six years later BAS launched the third edition, *An Environmental Agenda for Belize 2008-2013*. The same process as with the first two editions was used to generate a draft document. Then a Verification Workshop was held to gather broader input. Added subjects represent environmental issues that have risen to the top of our agenda: Environmental Governance, Climate Change, and Petroleum. *An Environmental Agenda for Belize 2008-2013* was launched on Earth Day, 2008, the theme of which was "Solutions to Climate Change". Many of the recommendations in the Environmental Agenda would help Belize adapt to any changes in our environment that might come along with climate change. Both the Executive Summary and the Full Reports were published and have been in great demand.

(*top*) BAS President David Craig presents the Environmental Agenda 2002 and Beyond to Hon. John Briceño (*bottom*) BAS Executive Director Anna D. Hoare presents the Environmental Agenda 2008–2013 to Hon. Gaspar Vega, Deputy Prime Minister.

Chalillo Hydroelectric Dam

In 1998 the GOB and Belize Electricity Limited (BEL) proposed to build a second hydroelectric dam to generate electricity. BAS, realizing that it had been nine years since the last analysis, began an initial consideration of options for Belize's future energy needs. Discussions were held with a number of knowledgeable people and a literature survey provided the background for a working paper. Alternative energy sources considered include co-generation with the sugar, citrus, and sawmill industries, along with wind, solar and diesel-generated energy. The Society reported the results in the "Working Document on the Renewable Energy Generation Options for Belize." The Society recommended that the GOB:

- formulate a long-term, comprehensive **energy policy** for Belize to reduce its dependence on the importation of fossil fuels and energy,
- take into consideration the proper and responsible use of the natural resources of Belize,
- weigh all the options available for generating power,
- conduct a vigorous and comprehensive review of **all** the options for generating energy from renewable sources, and
- conduct EIAs for whichever options are selected.

BAS continued addressing energy issues with the April 2001 special issue of the *BAS Newsletter* on "Renewable Energy", including a comprehensive review of potential energy options for Belize, a comment on the energy implications for climate change, a statement from the Natural Resources Defence Council, an appeal based on the area that would be inundated by the dam, an interview with the CEO of Belize Electricity Ltd., and a review of the Comprehensive Report from the World Commission on Dams.

Review of the Chalillo Environmental Impact Assessment

When the second Environmental Impact Assessment (EIA) became available in September of 2001, BAS commissioned several experts to review the various sections of the 1500-page EIA. The BAS Board of Directors and Advocacy Committee spent countless hours reviewing the expert's reports and discussing the BAS position. Initially the decision was that the Society could not support the Chalillo project because of important information gaps in the EIA.

- deficiencies in the EIA,
- inadequate public participation,
- watershed management,
- disaster management,
- inadequacy of hydrologic data,
- geologic concerns,
- effects on downstream wildlife,
- effects on the fishing industry,
- no analysis of the impact of the existing Mollejon Dam,
- no information on the planned third dam,
- no consideration of regional or global initiatives,
- only one alternative site considered, and
- lack of mitigation for archeological sites in the area to be inundated.

As the Association of National Development Agencies (ANDA) representative on the National Environmental Appraisal Committee (NEAC), BAS became part of the team of technical experts who analysed in depth each section of the EIA. The developers (BEL, BECOL, and Fortis) were cooperative in supplying the technical information requested by NEAC members, and as a consequence BAS was satisfied that its initial concerns had been addressed sufficiently to support environmental clearance.

But BAS still considered the consultation process was not extensive enough as it was largely restricted to the Cayo district. The Society is in full support of NEAC's recommendation for BEL to pursue a program of public hearings countrywide.

In order to ensure that the developers take the necessary measures to reduce environmental impacts, BAS through ANDA, will be fully involved in the formulation of the Environmental Compliance Plan, which is a legally binding document that incorporates the environmental mitigation measures identified in the EIA, as well as others recommended by the NEAC. BAS' conditional approval of the Chalillo Dam project was based on five conditions:

1) Performance Bond after Construction, which was completed.
2) Emergency Management Mechanism, which has not been completed. No early warning system has been implemented and there have been no evacuation exercises.
3) Complete Hydrological Monitoring is ongoing with the Hydrology Department.
4) Monitoring of Biodiversity, which is ongoing.
5) Continuous Public Information Sessions which was completed. Public hearings and information sessions were held with several consultations countrywide on the project.

The Macal Upper River Hydro-facility came on stream in November 2005. A third dam, the Vaca Dam was under construction in 2009. BAS monitoring has largely been through the NEAC and following up on concerns received through the general public. As the lead agency responsible for monitoring, DOE conducts regular monitoring of the site. BAS occasionally receives information through the DOE.

Land Policy Reform

An Advocacy Technical Committee was appointed by the BAS Board of Directors in early 1999. The focus of this committee for the period 1999 to 2001 was on land policy reform. The initial campaign was the gathering of alliances with all the stakeholders, including SPEAR and Belize Agro-producers Organization, to form the Land Alliance for National Development (LAND). The goal of LAND is:

> *...to influence the establishment, through legislation, and after full consultation with civil society and the private sector, a comprehensive, national land policy which allows for clear, non-partisan and fair practices for the distribution, management and sustainable use of our land resources ... no later than September 2001.*

Spearheaded by BAS, LAND developed a Draft Land Policy, which was presented at a national policy workshop on the 25th of July, 2001. LAND also lobbied for the establishment of the National Lands Advisory Committee and published *Survey of Land-Related Laws of Belize,* a guide to the laws of Belize the govern land ownership, development, use, and protection.

LAND National Policy Workshop in 2001

TILAPIA IN CROOKED TREE LAGOON

In May of 1996 it was reported that strange fish had invaded the Crooked Tree Lagoon and were taking over the lagoon, out-competing the native fish. BAS and the Fisheries Department collected samples from the Crooked Tree Lagoon and confirmed the presence of an exotic African species of fish of the genus *Tilapia* that was being farmed at Cherak's Farm at Big Creek. BAS began to release weekly press releases. In a Letter to the Editor in the *Reporter*, BAS Members Bruce and Carolyn Miller explained that *Tilapia* are very aggressive fish that reproduce large numbers of offspring rapidly. They can have a devastating effect upon ecosystems and native freshwater fish and may even affect the supply of lobster and shrimp in the sea by eating larvae at the shoreline.

This media attention resulted in an invitation by the Minister of Agriculture and Fisheries, Hon. Russell Garcia, for BAS to meet at Cherak's Farm. Several concerns were raised. No Environmental Impact Assessment was done until one year after the farm was in operation. There were no monitoring reports. There was a conflict of interest because those responsible for monitoring were personnel hired and paid by the Cherak's Farm. The screens in the catchment system were collapsed, resulting in water from the ponds flowing freely into the environment.

BAS and the Ministry of Agriculture and Fisheries agreed to form a joint team to address the concerns of the fishermen about *Tilapia* in the Crooked Tree Lagoon, to review the policies and regulations governing aquaculture in Belize, to ensure that all aquaculture project proposals go through an environmental impact assessment process prior to implementation, and to promote public awareness of the real economic, social and environmental effects of aquaculture in Belize.

BAS visiting Fresh Catch *Tilapia* farm along with Fisheries and Forest Department personnel. (*centre*) Tanya Williams-Thompson and Nellie Catzim investigating Tilapia farm.

Fish Kill in Crooked Tree Lagoon

Tanya Williams-Thompson

BAS responded to concerns raised by the Crooked Tree community over fish deaths in 2007 in the lagoon system of the Crooked Tree Wildlife Sanctuary.

Community members also raised concerns over the use of an electrofisher by researcher Peter Esselman to conduct research within the lagoon system. Esselman invited the community members to view a demonstration of his method. However, community members reiterated their opposition to the research method and refused to participate. BAS supported the immediate cessation of the research as a result of these concerns expressed by the community.

BAS sought the assistance of the Department of Environment (DOE), the Fisheries Department (FD), and the Belize Agricultural Health Authority (BAHA) to test water quality and fish samples for possible diseases. DOE notified the community that low dissolved oxygen was noted in areas of the lagoon where the fish deaths occurred. The lowest recorded level of dissolved oxygen was at .08 parts per millilitre of water (pp/ml). Tilapia, the most common fish found in the lagoon, needs a minimum of 2.25 - 3.0 pp/ml to survive.

Aquatic life such as fish require a certain level of dissolved oxygen. When that level drops, it creates conditions that can not support aquatic life. Low dissolved oxygen can be attributed to several factors, both natural and human-caused. This includes warmer water temperature which does not hold as much dissolved oxygen as colder water. Other causes could be bacterial growth that competes with aquatic life for oxygen, either naturally from decaying vegetation or from human activity on land that results in nutrients running off into the water.

BAS worked with the DOE to implement a monitoring programme and seek the community's involvement in conducting the programme.

Further concerns were raised about rumours that BAS was selling fish captured from the Crooked Tree Wildlife Sanctuary. BAS has investigated this issue and found no evidence to support this claim. BAS assured the community and all Belizeans that the organization would never participate in any activities that would have a negative impact on the resources upon which we all depend to sustain life and livelihoods. BAS is open to receiving any concerns from the public regarding any of the protected areas it co-manages and assures that full attention will be given to addressing it.

CTWS Director Donald Tillett (1985-2006) shows map of Sanctuary

Threats to Crooked Tree Wildlife Sanctuary

Tanya Williams-Thompson

In December 2004 leases on Black Creek within the Crooked Tree Wildlife were issued to developer Luke Espat and a road was cleared one and a half miles into the Sanctuary and within the 66-foot riverside reserve. The DOE lost its case against the developer and later settled out of court. The leases were never revoked.

The Belize Audubon Society (BAS) is the co-manager of the Crooked Tree Wildlife Sanctuary (CTWS) along with the Forest Department. On Thursday, April 3, 2008, the Belize Audubon Society was informed that a road was being constructed across the Western Lagoon within the Crooked Tree Wildlife Sanctuary as part of a development project to access farmlands and as part of a larger plan aimed at increasing tourism.

This was spearheaded by Hon. Edmund Castro and six villagers who provided some of their own funds to create a narrow filling within the lagoon. With the rains much of the causeway has been washed away and is now only passable during the dry season. This accessible pathway has also led to intruders entering the village and stealing cattle and ecologically it has impacted the waterways as it is blocked the free flow of the water in the Western Lagoon.

BAS is supportive of any sound infrastructural development project in improving the economic conditions of our communities, particularly for Crooked Tree Village where BAS has invested and implemented several community development initiatives. However, this infrastructural development has proceeded without going through the proper legal process and is contrary to the Laws of Belize that govern protected areas and development.

Development by Luke Espat within the Crooked Tree Wildlife Sanctuary on Black Creek. One and one-half miles of road (*top*) was cleared in the Sanctuary, even down to the riverside (*bottom*).

Road across Western Lagoon in Crooked Tree Wildlife Sanctuary built in 2008

Belize Audubon Society's position is that all projects should adhere to the laws of Belize, particularly when the development is planned for a sensitive protected area. Adherence to the laws would enable proper planning, reducing potential negative environmental, social/cultural and economic impacts.

The CTWS is a nationally and internationally recognized protected wetland area. International recognition comes from its designation as a Ramsar site which gives it the status of a wetland of international importance. While these recognitions are commendable, it is the valuable functions of the lagoons and creeks within the CTWS that makes it an important area.

The CTWS is the primary water supply source for the Crooked Tree Community and also provides food for the community through subsistence fishing. In addition, the lagoons and creeks play a critical role for flood control for the Belize River Valley and for Belize City as they act as a huge water storage area for the Belize River during heavy rains. This function was clearly evident in the aftermath of recent hurricanes.

People are not the only ones dependent on the water within the CTWS as the Sanctuary provides for critical habitat for wildlife with thousands of migratory and local water birds, including the Jabiru and Wood Storks, flocking to the CTWS to feed on fish found in the lagoons. It is this role that has attracted visitors to this protected area.

These functions must be safeguarded and taken into consideration in planning development in close proximity to the CTWS and even more so for a development in a wildlife sanctuary.

Cattle Farming Best Practices for Crooked Tree Wetland Conservation

Cattle Farmers, Agriculture Department and Belize Audubon are working together to keep Crooked Tree Wildlife Sanctuary "laik dat"! Thirty-one farmers from Crooked Tree, Lemonal, Gardenia, and Biscayne attended a Cattle Farming Best Practice session in 2008 organized by BAS and skillfully implemented by the Agriculture Department's Extension Officer Manuel Cowo.

The BAS Education programme implemented demonstrations on the ten practices developed in collaboration with the agriculture department:

1) Fencing of cattle to avoid loss of cattle and loss of energy as well as to keep them off the wetlands and the community roads,
2) Pasture Management (growing grass) to supply feed for cattle in the contained areas,
3) Leaving 300 feet or more (500 feet) of forest cover around waterways to avoid losing land or land erosion as well as to keep the role of trees in filtering pollution,
4) Leaving primary canopy (e.g. Ceiba trees) in farms for shade, water retention and Jabiru nesting,
5) Using trees such as Madre Cacao or gumbo limbo for fencing, to imporve soil conditions and reduce logwood cutting,
6) Using legumes or adequate types fertilizers that do not harm the environment,
7) Using some type of supplement feeding for cattle – sugar cane, hay, molases blocks etc.,
8) Creative and profitable use of cattle dung, so it does not run off into wetlands – vegetable gardens, composing and use in biogas,
9) Keeps animals contained during the dry season for the majority of the time to reduce soil compaction and grazing on the wetlands,
10) Keeps records of their business activity, and
11) Treat cattle with appropriate vaccines and medication.

These practices were promoted in the form of a comic booklet, a DVD showing implementation by farmers and explaining methodologies, a detailed booklet, cattle farmers' demonstrations, and a cattle farmer's forum.

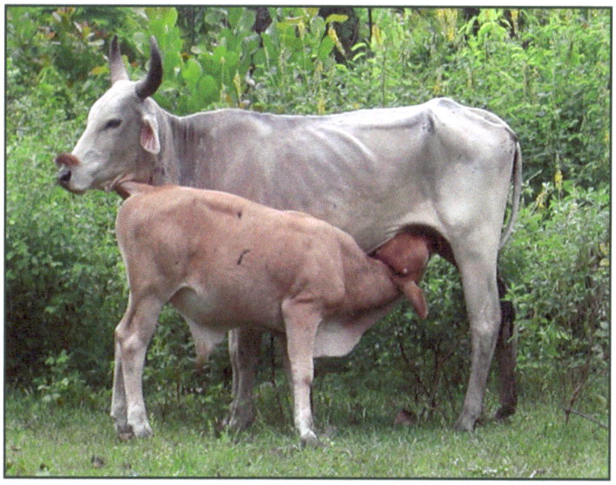

Cattle in Crooked Tree Village

(*top*) Manuel Cowo conducting a workshop for cattle famers of Crooked Tree Wildlife Sanctuary buffer communities.

(*bottom*) Demostration of best practices for cattle farmers.

BAS on Mega-Developments

One issue that prompted the Lamanai Room Declaration II was the proliferation of planned large developments for Belize, especially in coastal areas. Of ten projects reviewed by NEAC in 2003, eight of them were for coastal areas. BAS developed the following policy on mega-developments.

BAS publicly states its extreme concern over recent occurrences of wanton disregard for environmental laws established to regulate development in Belize. Of particular concern to BAS are the Port of Belize-Carnival Cruise Terminal, Stake Bank-Royal Caribbean Cruise Terminal, and the Drowned Caye Development projects. ... BAS has never been against development as is clearly spelt out in our mission statement of 'creating a balance between people and the environment.' Our assertion is that development must be planned and must be environmentally, socially, and economically sustainable.

These projects are expansive with great potential to cause negative impacts to the environment, as well as possibly having great social implications. Therefore, it is critical that thorough environmental impact assessments are conducted to ensure that these impacts are addressed ...

When developers are allowed to ignore the law by not applying for the necessary permits or conducting the necessary impact assessments for short-term profits, it is the environment, social and economic future of the country that is being put at risk.

Ara Macao Project on Placencia Peninsula

The Ara Macao project proposed accommodations for nearly 13,000 people, a 67-acrre marina, two 9-hole golf courses, and a 100,000 square-foot casino on the Placencia Peninsula. Residents organized to oppose this and several other projects proposed for the narrow 15-mile peninsula.

BAS voted against this project at the NEAC based on three key issues: 1) no adequate measures to reduce shore erosion, 2) size of the project has potential negative social and environmental impacts, 3) cumulative impacts of these high density projects has not been studied.

Drowned Caye/Stake Bank Developments

For both the Drowned Caye and Stake Bank Developments work has continued without the required permits or environmental clearance from DOE. Dredging and land reclamation were conducted without a permit in the Stake Bank and Drowned Caye areas. A causeway is planned that would connect these two projects to each other and to the mainland.

The Belize City area, including both Stake Bank and the Drowned Cayes, are so important to the endangered manatee that in 2002 the Swallow Caye Wildlife Sanctuary was created to protect this critical habitat for the manatees. Scientific studies have shown that large population of manatees utilize the area between Northern Drowned Caye, Swallow Caye and Stake Bank.

As a representative member of the NEAC, BAS voted against the Stake Bank project. However, this project was approved by the majority of the NEAC members.

Port of Belize-Carnival Cruise Terminal

Land reclamation activities for the Port of Belize-Carnival Cruise Terminal project have been underway since 2002, despite the facts that no EIA was developed until 2007 and DOE issued a stop order. Furthermore, there is concern that this project is negatively impacting the nearby communities.

South Beach Development on Ambergris Caye

AMANDA BURGOS ACOSTA

BAS does not support the mega project South Beach Belize for various reasons, most notable being:

- in this particular case, the proximity of Hol Chan Marine Reserve to the proposed development places the project in a sensitive area, making it imperative that any development be approached with caution;
- the infrastructural demands of these developments cannot be met given the current systems in place (such as solid waste disposal and labour force); and
- given the global economic climate, all financial inclinations are that investments in second home mortgages and vacation properties are decreasing in viability, increasing the likelihood that mega projects may be left unfinished.

Spanish Bay Development Project

TANYA WILLIAMS-THOMPSON

The Spanish Bay Project was a proposal to further develop Spanish Lookout Caye to include recreational tourism, education, and research components. The recreational tourism section included a captive dolphin park, which was the major concern about this project. BAS stated that "the keeping of dolphins in captivity is incongruent with conservation efforts in Belize, as well as creating an erroneous perception of dolphin behaviour in the wild. ... BAS' view is that such programmes encourage the capture of wild dolphins, an inhumane and unsustainable practice."

Protecting Belize's Water Resources

TANYA WILLIAMS-THOMPSON

BAS sits on the National Pro Tempore Water Commission and has been lobbying for the endorsement and implementation of the "Integrated Water Policy and Water Resources Legislation for Belize" which was presented to public consultations on June 13, 2006.

Waste Management

DIANE WADE-MOORE

Waste management in Belize has suffered from indecision. Every edition of the BAS Environmental Agenda has a section devoted to waste management. The site close to Belize City is a health hazard and must be closed, but lack of agreement on location of an alternate new site has stalled progress. BAS has lobbied for an end to the indecision.

Earth Summit

In preparation for the Earth Summit held in Brazil in June, 1992, BAS coordinated the Belize Summit on Environment and Development. Eight priorities were discussed in concurrent workshops: regional cooperation and institutional aspects; information, communication, education and training; research, creation and transfer of technology; planning and management of the environment; general policies; economic and financial aspects; control of pollution; and legislation. Belize joined the Central American Commission for Environment and Development (CCAD) in their effort to arrive at a Central American position with respect to the mode of development that is most appropriate for the region.

Oil Refinery Planned for Monkey River

In 1971 BAS became aware of a proposal to put a $150-million oil refinery on the southern coast near Monkey River and expressed concern.

> *While welcoming an advance of such magnitude in our economic growth, our society is apprehensive of the possible danger of pollution which an oil refinery might bring to beaches and waters with consequent loss in tourist business, loss to our fishing industry, and, more particularly, the threat to our pleasant quality of life. And mindful of the fact that prevention is better than cure, and far easier, we would therefore urge that sufficient covenants with stiff penalties should be written into any contract granting a license to build and operate the proposed oil refinery. We suggest particularly that industrial waste shall not be dumped into the sea, that the licensee clean up any oil spills and be liable for any breach. (Letter to Minister of Trade and Industry, 13 November 1971)*

We wrote to conservation organizations in the U.S. for information, which was passed on to the responsible persons in Government. This went on for some time, but the project was eventually cancelled.

Oil Exploration in Sarstoon-Temash National Park

The Sarstoon Temash Institute for Indigenous Management (SATIIM), the co-managers of the Sarstoon-Temash National Park, were informed that US Capital Energy was ready to commence seismic testing for oil within the Sarstoon-Temash National Park. SATIIM sued and got an injunction in 2006 against the seismic testing because no Environmental Impact Assessment (EIA) was done. Once the EIA was completed, the seismic testing was approved and commenced in April of 2009. SATIIM's lawyers argued that oil exploration in the National Park violated the co-management agreement between the GOB and SATIIM, but the court ruled that the co-management agreement had no legal basis in the Laws of Belize. However, the court also ruled that while seismic testing did not violate the National Parks System Act, drilling for oil in a national park would violate the Act.

As a result, BAS developed a position statement on oil exploration in national parks:

> ... BAS recognizes that the budding oil industry in Belize cannot contribute to the sustainable economic development of the nation without proper adherence to environmental and socioeconomic policies and plans. BAS also recognizes that these national policies and plans will be flawed without full, transparent stakeholder consultation, participation and consensus. Learning from the mistakes of countries worldwide regarding the oil issues, priority should be given to policy development for oil exploration and production to ensure best environmental practices and equitable distribution of the benefits derived from oil production. This will guarantee that Belizean citizens continue to maintain the basic, universal right of access to environmental services such clean water, air and healthy environment. ...

BAS is also cognizant that the extraction of oil could have devastating impacts upon the ecosystems and biodiversity of any protected area. Oil exploration significantly changes the ecological character of the environment and may cause detrimental and irreversible damage to the natural environment. Belize has joined the nations around the world that declare national lands as protected areas to ensure the ecological integrity and perpetuation of natural resources and biodiversity, while contributing to sustainable development.

Tourism—the second most significant contributor to Belize's social and economic development—is dependent on the integrity of its protected areas. The National Protected Areas Policy and System Plan recently endorsed by Government, aims to safeguard the natural resources found within protected areas. However, the integrity of these natural resources can be degraded by carelessly implemented development projects or projects that do not address sound environmental management.

BAS also believes that oil exploration within the marine areas should have extremely stringent conditions applied due to the high vulnerability of the marine ecosystems. Oil exploration should be excluded from the protected marine areas such as the Belize Barrier Reef and the three atolls that are invaluable to the tourism and fisheries sectors, and for protection of coastal communities. Oil exploration could place the Belizean reefs and coastal areas under unacceptable risks.

As a protected areas manager, BAS does not support any activity that would compromise the ability of our natural systems to maintain a healthy environment that is critical to sustaining the Belizean population and economy. Belize Audubon Society, as embodied in our mission statement, has a responsibility to work with all stakeholders, at all levels and in all sectors, to try to achieve that balance between the needs of the people and the environment.

Judy Lumb

BAS Executive Director signing the poster, joining the Belize Coalition ot Save Our National Heritage.

Offshore Drilling for Oil

A coalition of organizations was formed to lobby for a ban on offshore drilling for oil in Belize. The Belize Coalition to Save Our National Heritage was launched on World Ocean Day, June 8, 2010. Representatives of organizations signed a poster to indicate their participation in the Coalition. Melanie McField spoke on the environmental impact of offshore oil; Eddi Romero spoke on the economic value of the reef for protection from storms, support for the fishing industry, and tourism, in contrast to the potential economic value of oil; and Audrey Matura-Shepherd spoke about the legal realities of oil concessions in Belize. She emphasized that the laws of Belize provide no protection against damage from oil exploration or spills. Class action suits are illegal in Belize. Only the GOB has the right to extract oil. Under the Petroleum Act the Minister makes all decisions. There is no provision for consultation or an independent regulatory body. There is much work to be done.

BAS Advocacy Programme Manager Tanya Williams-Thompson (*center*) and BAS Executive Director Anna D. Hoare (*second from right*) on KREM morning show.

BAS Media Campaigns

The media continues to be a critical tool in forwarding advocacy initiatives and addressing concerns, so BAS continues to work with the media to enable sustained awareness on key issues.

A national environmental public awareness campaign, sponsored by a grant from PACT, was conducted from June 1998 through February 1999 to promote Belize's protected areas. The hope was that more Belizeans would visit these natural areas and, hence, develop a greater awareness and appreciation for Belize's natural resources and overall environment. Six video TV spots were prepared and broadcast on local television stations. Similar information was prepared for radio and newspapers throughout the country. Brochures about Belize's conservation laws were developed for immigrants and for owners of large parcels of land in Belize.

Between 2007 and 2009, BAS appeared in more than thirty talk shows and news programs. Here are a few of them:

- Unsustainable Development/North Drowned Caye & Stake Bank Caye – *Channel 7*
- Crooked Tree Fish Deaths – *Channel 5*
- Coastal Development – *CTV 3*
- Lamanai Declaration and BAS Advocacy Programme – *KREM*
- Oil Exploration – *KREM*
- Climate Change – *Channel Five*
- Nature and People – *Channel Five*
- Bird Conservation – *Belize Watch, Love FM*
- Co-management – *Love FM/Channel Five*
- Environmental Issues – Plus TV
- Community Development – Belize Watch

40 YEARS OF BAS ADVOCACY

Advocacy for wildlife, environmental quality, and sustainability of life in Belize has been a part of the work of BAS since the first efforts to protect the Jabiru. Early BAS efforts included lobbying against such potentially destructive projects as an oil refinery on the coast near Monkey River or a game ranch with exotic animals near Crooked Tree. BAS was involved in the development of the initial legislation to protect wildlife and establish a system of protected areas, and every piece of legislation and policy affecting protected areas and the quality of life in Belize ever since. In this way BAS has helped to keep Belize as pristine as it is today.

BAS has been active in protecting Belize's spectacular and extensive marine resources. Through the 1980s and 1990s, the BAS Reef Preservation Committee was active in the management of Half Moon Caye Natural Monument and the protection of sea turtles. As management of Half Moon Caye and Blue Hole Natural Monuments was developed, advocacy for marine issues was taken on by BAS Marine Protected Areas Staff.

Thanks to funding from HIVOS and other sources, excellent staff have been available to carry out a great variety of advocacy projects, from public education campaigns to reviews of development projects at NEAC to support for BAS-managed protected areas and their buffer communities. Sitting on a number of national committees, BAS is able to add the long-range view toward sustainability and environmental quality to any decision-making process.

In 1997 BAS convened a joint Board meeting with the Belize Tourism Industry Association and invited other environmental organizations to resist uncontrolled, environmentally destructive development. The result was the first Lamanai Room Declaration. Ten years later BAS was concerned about even greater issues with uncontrolled, unsustainable development without proper planning, so a second gathering of environmental and tourism organizations was convened, resulting in Lamanai Declaration II.

For the past three election cycles, BAS has developed a major advocacy instrument, *Environmental Agenda,* which presents the basic issues that need addressing as the political parties develop their manifestos. All three of these documents published in 1998, 2002, and 2008, have been influential in the political arena.

The most controversial project was the proposed Chalillo Hydroelectric Dam project. BAS spent considerable effort studying and gave it conditional support as a low carbon source of electricity.

BAS led the way in the development of a campaign for land policy reform. The Land Alliance for National Development (LAND) was formed with BAS serving as the Secretariat.

But the work described in this chapter covers only the major BAS advocacy projects. Not included are the small services performed on a daily basis by BAS. The public views BAS as the ultimate authority when it comes to any issue relating to wildlife. "Come get this snake out of my bathroom!" is a typical request. It is not BAS, but the Forest Department that is the ultimate authority in regard to wildlife in Belize. Still, BAS addresses all requests as best it can, given limits of staff and resources.

The work of the Advocacy Programme at BAS has made major contributions to the development of Belize, encouraging sustainable development while trying to reign in the unplanned, unsustainable development that is increasing in Belize at the time of the BAS 40th anniversary.

Belize Audubon Society

Celebrating 40 Years

Chapter Three

Environmental Education

Goals and Objectives of Environmental Education

Anna D. Hoare

The overall goal of Environmental Education is to change peoples' behaviour to:

- adopt more environmentally appropriate behaviour and proper stewardship of resources,
- develop new attitudes and behaviours that help sustain life for today and future generations,
- improve quality of life of the citizenry of Belize,
- recognize how resources can be used more efficiently, and
- identify alternatives to destructive development practices which destroy the environment and affect the livelihood of people.

Youth Education Committee

Lydia Waight

From its inception the Belize Audubon Society has been committed to environmental education, both of youth and the general public. The Youth Education Committee was established at its formation and members quickly got to work. Copies of the *BAS Bulletin* were provided to all secondary schools. BAS members were invited to present lectures at a number of different organizations including the Trade Union Seminar for Worker's Education and the National 4-H Camp at Gracey Rock.

At the monthly meetings of the Society there were talks, lectures, slide presentations and movies by members and others sharing their expertise. Business meetings usually followed the feature, which was advertised as being open to the public. The talks were given in 1969 are listed in the table following.

Talks at BAS Meetings in 1969		
March 7	Dora Weyer	Birds of Belize City
April 17	Magnus Halcrow	Orchids of Belize
May 22	W. Ford Young	Physical Geography and Geology of Belize
June 19	Meg Craig	Tropical Plants
July 24	Louis Lindo	Forest Environment
August 21	Peter Hill	Movies of Wasps, Butterflies and Snakes
September 12	Dr. Andre Clewell	Use of Fire in Forest Management
September 18	Gil Rosado	Our Fishery Problems
October 23	Dora Weyer	Snakes of Belize (with live specimens)
November 20	Peter Hill	Movies of Birds, Mammals, and Insects

Our December meeting was cancelled as a mark of respect for Russell Waters, Chief Forest Officer and BAS member, who was killed in a plane crash on December 8, 1969. He had been scheduled to speak at the January 22, 1970, meeting. Instead Magnus Halcrow spoke on "Conservation in West Africa."

Radio Programmes

Beginning in July of 1969, BAS radio programmes were written and presented by members as a part of the Scout Programme on Radio Belize. Board member Norman Staines was Executive Director of the Scouts and arranged for BAS to include these scripts on natural history. Subsequently Radio Belize donated airtime for "Audubon Weekly" read on Mondays. These radio scripts were printed by BAS for distribution to schools and other interested parties. Visitors were interviewed and on one programme student members participated in a panel discussion.

Summer Programme in Tropical Biology

In 1970 a group of U.S. mid-western universities formed the Associated Universities for International Education and established a Tropical Research Centre at St. John's College. Courses in marine and terrestrial ecology were offered from 1972 through 1977. The marine course was held at Glovers Reef and the terrestrial course at various locations on the Belize mainland. Mrs. Weyer was on the faculty and scholarships were offered to Belizean students. BAS took the opportunity of sponsoring public lectures by faculty members at the Bliss Institute on a wide variety of subjects.

Goshen College Programme

From 1974 until 1979 BAS conducted a cooperative environmental education project with Goshen College in Goshen, Indiana. Goshen students came to Belize for their Study/Service Trimester. Some worked in protected areas and others presented the BAS Nature Study Programme to schools. BAS members prepared slide presentations with taped narrative and the Goshen students took them to the schools in Belize City and in the Districts. The first year, 1975, more than 3,700 Belizean students saw the two presentations, "The Coastal Edge—Sand Beach" and "The Coastal Mangrove Edge." In 1978 the 72-slide presentation on "Hardwood Forests" was shown to 6,900 students.

Obtaining equipment for these projects was always a challenge. The Goshen students brought the first slide projector, which was purchased with BAS funds. Then an offer came for another donation of a tape recorder and a screen from *Centro Agronomico Tropical de Investigacion y Enseñanza* (CATIE).

BAS Library

The first BAS library was housed at Ford Young's office and was comprised of his books, which he donated. Additional books were donated by Colonel Hastings and other members and friends. The library was moved to the BAS office on the Foreshore when it was opened in 1985. Mrs. Catherine Meekin, a Peace Corps Volunteer, spent a few months cataloguing the books and made a card index for the *BAS Bulletin*. By a cooperative arrangement in 2000 much of the BAS Library was moved to the University of Belize to allow greater access.

BAS Bulletin

The first *Belize Audubon Society Bulletin* was produced in March 1969 by W. Ford Young and his wife, Callie. In September Norman Staines succeeded them and then the BAS Secretary, Mrs. Lydia Waight took on the task by the end of that first year. She faithfully produced the *BAS Bulletin* every month for the next seventeen years with only a typewriter and mimeograph machine. In August of 1987 the *BAS Bulletin* was succeeded by the quarterly *BAS Newsletter.*

Valdemar Andrade teaches in the Tapir Mountain Environmental Carnival.

Coastal Zone Management Project

June Neal and Valdemar Andrade

Beginning in March of 1990, a BAS staff member was responsible for the Environmental Education component of the Coastal Zone Management (CZM) Project under the Fisheries Department in the Ministry of Agriculture and Fisheries.

It was recognized that the country needed to adopt a multi-sectoral approach in developing a long-term strategy for coastal resources management and conservation, which would complement the rapidly expanding economic development of the country. The major focus of the CZM project was the preservation and sustainable use of the Belize Barrier Reef, the longest barrier reef in the Western Hemisphere. The variety of reef formations occurring within the territorial limits of Belize is unparalleled elsewhere in the Caribbean. The task of managing this system on a sustainable basis is an enormous challenge and one that warrants extraordinary measures. A strong emphasis was placed on education, at all levels, on coastal issues. June Neal was the first Environmental Educator under this programme followed by Valdemar Andrade in 1993.

Rainbow the Toucan at the Wesley Primary School in Belize City. Osmany Salas is inside the costume.

ENVIRONMENTAL CONSERVATION CARNIVALS

Two travelling environmental carnivals used Rainbow the Toucan and other innovative techniques to bring environmental education to students of Belize's outlying districts. Environmental educators from the Belize Audubon/Coastal Zone Management Unit, the Belize Centre for Environmental Studies and the Belize Zoo presented the Environmental Conservation Carnival to all the primary schools in the Toledo District in the Spring of 1995. In the Fall, the carnival was taken to all the schools in the Corozal District. In October the presenters made a special appearance at schools in the villages near Tapir Mountain Nature Reserve in the Cayo District. This ambitious project was funded by a flock of agencies including World Wildlife Fund, the Forestry Planning and Management Project and the Natural Resources Management Project (NARMAP).

RARE PROGRAMME

The RARE (Rare Animal Relief Effort) of the Center for Tropical Conservation of Philadelphia, Pennsylvania, was an environmental education project that was carried out by the Forest Department of the Ministry of Natural Resources to promote public awareness. Beginning in the Spring of 1993, BAS assisted with Belize City schools where the RARE programme was presented by the CZM Environmental Educator and BAS Staff. In one month they visited some 43 schools and met over 10,000 students. Mass media and puppet shows were used, including the human-sized Rainbow the Toucan mascot. The programme was a hit with students because it made learning about the environment entertaining as well as educational.

Students of Xavier Primary School with their teacher at the Corozal Environmental Carnival held at the Corozal Civic Center

Training Programmes

During our 40-year history BAS has participated in many training programmes. Taking advantage of support available from the Small Grants Programme of the Global Environment Facility implemented by the United Nations Development Programme (GEF SGP), the World Conservation Union (IUCN), United States Agency for International Development (USAID), World Wildlife Fund (WWF), National Resources Management Programme (NARMAP), which was supported by USAID and implemented by WWF, and other sources, training programmes have focused on increasing the capacity of the buffer community members, BAS staff, and teachers.

Tour Guide Training

In 1987 BAS assisted the Belize Tourism Industry Association in providing local tour guides with knowledge of the natural history of Belize, major ecological zones, flora and fauna, and conservation.

Economics of Natural Resources

In 1993 "Economics of Natural Resources" was the subject of a Seminar coordinated by BAS and sponsored by NARMAP, the MacArthur Foundation, and the *Universidad de Concepcion* of Chile. Dr. Arcadio Cerda of the *Universidad de Concepcion* of Chile was the main facilitator.

Conservation and Management Training

BAS received a grant from GEF SGP for the training of protected areas managers and local community leaders in 1994, with fieldwork in Guanacaste National Park. The fourteen participants represented six protected areas and four buffer zone communities. A follow-up workshop was held at Shipstern Nature Reserve with the same participants.

Gender and Development

Sponsored by IUCN, BAS hosted a workshop entitled "Gender and Development" at the Pelican Beach Resort in 1994. Lorena Aguilar Revelo from IUCN's Costa Rica office did a fantastic job facilitating the workshop, which was designed to increase the awareness of societal attitudes and values concerning gender and how these attitudes change over time, to show experiences and case studies, and to show how to integrate gender and development in an organization. Participants included representatives of women's organizations, environmental organizations, buffer zone community groups, and BAS protected areas staff.

BAS Archives

Participants in the Gender Workshop

Freshwater Creek Training Programme

The Association of Friends of Freshwater Creek, which was formed to co-manage Freshwater Creek Forest Reserve, the only declared Forest Reserve in the entire northern Belize. The Association is comprised of members from Orange Walk Town, San Estevan, Santa Martha, and Chunox, who approached BAS for assistance in attaining the technical expertise and management skills that are needed for co-management.

BAS Archives

Association of Friends of Freshwater Creek on a field trip to Cockscomb Basin Wildlife Sanctuary with Michael Somerville on the far right.

Environmental Education Programme

Michael Somerville, 1998

A variety of Environmental Education efforts are underway in Belize by both the public and private sectors. Programmes are working to increase the school children's awareness and understanding of environmental problems, developing resource materials for teachers, and increasing public awareness about various conservation and environmental issues.

A baseline study was conducted in an effort to develop an environmental education strategy for the country. It was concluded that the most important audiences for addressing and resolving environmental problems are educators, NGOs, natural resource managers, community leaders, government officials, farming communities, environmental specialists, children, and media personnel.

While past environmental education programmes proved effective to some extent, most if not all of them have been confined to short-term periods, mainly as a result of the lack of adequate financial resources and personnel. BAS recognizes that in order for any environmental education programme to be truly effective, it must be continuous.

The BAS Environmental Education Programme is aimed at all levels of the general population on the premise that educating people about the environment and conservation will reciprocally lead to a better way of life for now and for generations to come.

On-the-road exhibits have been developed to reach residents of buffer zone communities and the general public, to educate them about Belize's protected areas, the overall environment, and good conservation practices.

Training for BAS Staff

Over the years many training programmes have been held to increase BAS institutional capacity. Staff have acquired skills in Project Planning, Management, Evaluation and Log-frames, Rescue Training, Climate Change, Accountability and Good Governance, Green Laws and Enforcement, Special Constable, Security and Safety, Contingency Risk Planning and Hurricane Preparedness, Community Rural Tourism, Coral Reef Ecology, Statistical Analysis and Hypotheses, Database Management, Strategic Environment Assessment, Leadership Training, Bird Monitoring, Boat Handling and Safety, Video Production, Environmental Interpretation, Media communication and Advocacy, Communication for Supervisors, Marketing, Team Building, Organisational Development, Eco-regional Planning and Fire Management.

In the Summer of 1989 Dr. Douglas James gave a series of two-day workshops to personnel of CBWS, GNP, and BHNP on the techniques of bird identification funded by the US Fulbright Programme.

Nine BAS wardens were trained at the Police Training School in Belmopan and sworn in as Special Constables in 1991. In 1992 three additional wardens were trained as Special Constables and, in 2004, fourteen wardens were given two weeks of intensive police training and achieved the status of Special Constables.

In 1991 a course was held at the Douglas D'Silva Station in the Mountain Pine Ridge Forest Reserve, organized by the Forest Department and Colorado State University. CBWS Park Director and Protected Areas Manager Osmany Salas were Assistant Facilitators.

Special Constable training: Brijilio Bolon receives his Special Constable certificate in 2004.

BAS wardens receiving first aid and CPR training

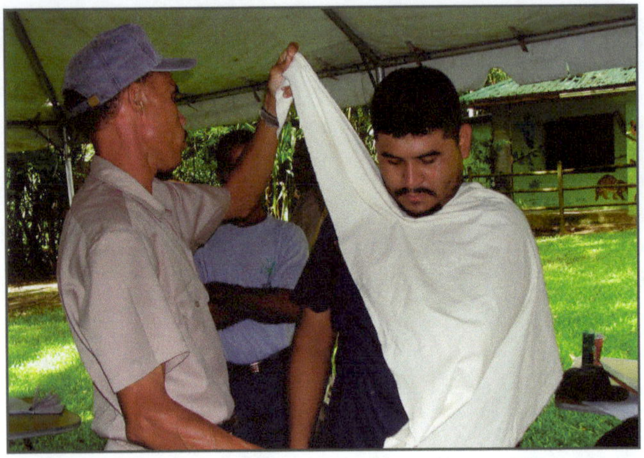

Uriah Rhaburn practicing first aid on Heron Moreno.

Derick Hendy learning to band birds.

In August of 1992 a training programme in wildlife management was held at the Belize Zoo and Tropical Education Centre, taught by Dr. Jack Meyer from the University College of Belize.

A minicourse on Photogrammetry and Rapid Ecological Assessment (REA) was given at the Pelican Beach Resort in Dangriga in 1993. The information was quickly used in a REA of the Bladen Branch Nature Reserve conducted by Pio Saqui, Eugenio Ah, and Donald Tillett.

In 2003 BAS wardens were training in basic first aid and cardiac-pulmonary resuscitation.

Four BAS wardens, Kent Garcia, Derick Hendy, Israel Manzanero, Jr. and Marcelo Pau, received training in bird banding from Dr. Nick Bayly and Camila Gomez as part of a research project entitled "Evaluating a Stepping Stone for Neotropical Migratory Birds – the Belizean North-East Biological Corridor."

BAS President David Craig and Education Manager Nimmi Herrera present certificate to Israel Manzanero, Jr.

Nature Schools

Olivia Carballo-Avilez, 2006

One of the most effective forms of environmental education is hands-on learning. Students particularly enjoy learning by exploring and experiencing their natural environment for themselves! Nevertheless, it is very important to be organized and prepared when conducting these activities as they can easily get out of control. BAS-managed protected areas are excellent "Nature Schools". For many years, teachers around the country have been using our protected areas. The most popular site with schools is Guanacaste National Park, followed by St. Herman's Blue Hole National Park, Cockscomb Basin Wildlife Sanctuary, Crooked Tree Wildlife Sanctuary and, occasionally, Half Moon Caye Natural Monument. To ensure that the experiential field activities at our protected areas are successful, BAS assists the organizers in coordinating their field activity.

The BAS team of field educators consists of our hard-working wardens at each protected area. They are very familiar with their natural environment and are excellent in interpreting the natural environment to students of all ages. The Education Programme Manager conducted short trainings to update them with information related to the specific protected area they manage—basic ecological principles; forest ecology (terrestrial protected areas); and coral reef, mangroves, and littoral forest ecology (marine protected areas). Interactive methodologies and strategies to engage students and large group management were also shared.

(*above*) Artist Bill Borella in the display he created in the St. Herman's Blue Hole National Park Visitor Centre.

(*below*) Cockscomb Basin Wildlife Sanctuary Park Director Nicacio Coc explains about jaguars to visiting children.

(*top*) Teachers learning to use Education Kits in 2001
(*centre and bottom*) Teacher's Day at Cockscomb in 2008

Training for Teachers

One way to influence environmental education of youth is participation in the development of curriculum and the training of teachers. In the 1970s the Goshen College programme included nature study and conservation in the syllabus for the Upper Division of the Primary Schools, Standards IV, V, and VI. After the Goshen programme was discontinued the teachers asked the BAS to continue presenting those nature study programmes.

BAS was invited to participate in the training of teachers at the Teachers College and participated in the Curriculum Development Unit's Infant Schools Science and Social Studies Seminar.

Education Kits

Environmental education kits filled with colourful and imaginative, hands-on activities were developed in 2001 by BAS member Anna D. Hoare and artist Bill Borella. Sponsored by the MacArthur Foundation, training sessions in the use of the kits were held for BAS field staff and teachers from buffer community schools.

Teacher's Day at Cockscomb

On May 15th, 2008, BAS hosted 75 children and 10 teachers from Maya Mopan to learn about Jaguars and their importance at Cockscomb Basin Wildlife Sanctuary. It was Teacher's Day so BAS staff supervised the 75 children and gave the teachers a break to go tubing and hiking. The group was divided into five groups and they all participated in jaguar education activities and ended the day with a jaguar drama, jaguar mask making, and face painting. The jaguar mascot came out to sing, play, and finally bid the teachers and children good-bye.

Masks and Cut-outs

These masks and cut outs were developed by volunteer David Rosenzweig in 2006. They are popular at the protected areas with visiting school groups, and used often in school presentations and decorations.

Wetlands Summer Camp

OLIVIA CARBALLO-AVILEZ

The 2009 Wetlands Summer Camp for Teachers was part of the Wetlands Campaign with the following theme "*Crooked Tree Wetlands ga wi baak; as we develop, mek we kip it laik dat!* (Crooked Tree Wetlands support us; as we develop, let's keep it intact!)"

At the Wetland Summer Camp CTWS staff talked to the teachers about the biodiversity of CTWS using the Visitor Center as the educational tool.

Cockscomb Basin Wildlife Sanctuary Staff hosted the teachers from the Wetlands Summer Camp for an overnight and talked to them about how Cockscomb is connected to Crooked Tree. Part of the Belize River Watershed starts from waterfalls in Cockscomb.

The teachers were so excited about the Wetlands Summer Camp that they wanted to share the knowledge they acquired with their community. So they turned the first part of their Parent Teachers meeting into "crooked tree wetlands ga wi baak" session!

(*top*) Teachers in the Wetlands Summer Camp
(centre) Olivia Carballo-Avilez conducting a wetlands demonstration
(*bottom*) Teacher Katie Gillett conducting her wetlands demonstration for parents.

Exhibitions

BAS takes every opportunity to educate the public on conservation, biodiversity, and the work of the Society. The table below lists some of the exhibitions in which BAS has participated with a display. Creating and setting up these displays requires considerable effort on the part of BAS staff, so the position of Public Participation and Publicity Coordinator was established in 2002 and Dirk Francisco took up this position.

Exhibitions		
Year	**Location**	**Subject**
1973–2008	Belmopan	Agriculture Show
1990	Bliss Institute	BAS Exhibition
1991	Bliss Institute	Feathers 'N Fins–Photo Exhibition
1992	Bliss Institute	Belize in Natural History–Photo Exhibition
1992	Bliss Institute	Mangroves: Our Coastal Zone
1994	Holy Redeemer	Eco-Fest–BAS 30th Anniversary
1994	San Ignacio Hotel	Hike and Bike for the Rainforest
1996	Bliss Institute	Turtle Week
1997	Holy Redeemer	Belize Barrier Reef
1998	Bliss Institute	Manatee Week
1999	St. Johns College	Our Reefs: Caribbean Connection
2000	ECPO 19	Rainforest Exhibit
2000	Cahal Pech Hotel	Sea Turtle Workshop
2002	Radisson Fort George Hotel	National Marine Forum
2002	EXPO 21	BAS display
2003	Mexican Institute/St. Johns College	Natural Heritage
2003–present	Cashew Fest	BAS display
2005–2008	Old Belize	BAS display
2005	Holy Redeemer	Natural Resource Week
2009	Mexican Institute, Benque Viejo, San Ignacio, Hopkins, Orange Walk, Corozal, Punta Gorda	Shutter Speed Photo Exhibition
2010	Crooked Tree Health Fair	BAS Display

BAS displays: (*above left*) Earth Day 2003 at CBWS, (*above right*) Musicians Titman Flores and Andy Palacio at EXPO 21 in 2002, (*right*) Natural Heritage photo exhibit at St. John's College in 2003, (*below left*) Dirk Francisco with BAS display at Natural Resource Week in 2005, (*below right*) Earth Day 2008 display at Guanacaste National Park

Judy Lumb

EARTH DAY CELEBRATIONS

Every year BAS celebrates Earth Day. Some examples include the 1994 Earth Day poster contest entitled "Spare the Earth – There Is No Spare."

1995

The 25th anniversary of Earth Day was celebrated in 1995 under the theme, "The Time Has Come" with a tree-planting ceremony at the U.S. Embassy in Belize City on April 21st and at the Unitedville Community Centre on April 22nd. The Saturday all-day family event in Unitedville focused on the three Cayo BAS-managed protected areas.

2000

A clean-up campaign celebrated the 30th anniversary of Earth Day in 2000 with 1500 students from ten Cayo primary schools followed by a ceremony at Guanacaste National Park.

2002

In 2002 Earth Day celebrations took place under the theme, "Protect Our Home" at Guanacaste National Park, Crooked Tree and Cockscomb Basin Wildlife Sanctuaries. Activities included roadside clean-up on the Western and Hummingbird Highways, essays, plays and songs created by primary school children.

2003

Teachers and students from the Cockscomb Basin Wildlife Sanctuary buffer communities came to the Sanctuary for a three-day jamboree for the 2003 Earth Day celebration under the theme, "Water for Life."

The theme of the three-day Earth Day Jamboree at CBWS in 2003 was Water for Life.

Earth Day 2006 was celebrated under the theme "Solutions for Climate Change" in Cockscomb Basin Wildlife Sanctuary (*above*) and St. Herman's Blue Hole National Park (*below*). Planting a mahogany tree at CBWS (*right*).

2005

BAS partnered with the Ministry of Natural Resources to expand the 2005 Earth Day to Natural Resources Week. Exhibits represented 24 protected areas and NGOs responsible for managing protected area, as well as five other NGOs, nine government agencies, and four funding agencies at the Holy Redeemer Parish Hall in Belize City from April 18 - 21.

2006

Supported by the Protected Areas Conservation Trust, "Solutions for Climate Change" was the theme for the 2006 Earth Day celebrations at St. Herman's Blue Hole and Guanacaste National Parks, Crooked Tree and Cockscomb Basin Wildlife Sanctuaries. BAS held events at each protected area for the children of the buffer communities that were entertaining while being educational.

2008

On Monday April 21, 2008 BAS celebrated Earth Day at Guanacaste National Park connecting Mother Nature with our golden citizens and students with special needs. BAS seeks to achieve the dream of having Guanacaste National Park accessible to everyone—the young, the elderly and the handicapped. It was great to see the excitement on everyone's face as they arrived at Guanacaste.

The students gave wonderful performances with their poem and dance. The poem "My Gift" written by the principal of Stella Maris School was presented by the students. Rowan Garel, our nine-year-old Junior Guide on the nature walk, identified about 12 birds by their calls. Even though he could not see, he made an excellent guide.

Leela Vernon gave a great performance having the students and elderly party a little. We then ended the day with a Toast to Mother Nature and a bit of dancing. We created a new partnership as we celebrated with two groups we often forget.

BAS launched a major "Greening Initiative" on Earth Day in 2008 with the slogan, "KEEP BELIZE GREEN, GO GREEN - IT MAKES ENE." Green bottles and green bags were distributed, highlighting the need to "Reduce, Reuse, and Recycle". A door hanger was developed for hotels to promote a "green life". One thousand mahogany trees were distributed.

My Gift

Thank you earth so green—so free
With your pretty face for all to see
My daily bread you provide for me
Because planting a seed is the key
Today I bring to you a gift—yes a tree
So that others too may cherish thee

Joy Westby
Principal, Stella Maris School

(*above*) Earth Day 2008 was devoted to the elderly and the handicapped. Stella Maris students presenting the poem, "My Gift". (*right*) Guanacaste National Park trails are made wheelchair accessible.

2009

As a part of the BAS 40th anniversary in 2009 BAS celebrated Earth Day with a "Walk for a Green Belize" from Belize City to Guanacaste National Park (*see Chapter One*).

2010

"Walk for a Green Belize" was such a success in 2009 that a bigger event was held in 2010 with 800 participants and 1,300 garbage bags were filled.

(above) Embassy of Mexico in Belize group participating in the "Walk for a Green Belize".
right top) Leela Vernon entertains and senior citizens dance at 2008 Earth Day celebration.

(*right bottom*) BAS Executive Director Anna D. Hoare presents a green bag to Hon. Gaspar Vega, Deputy Prime Minister.

World Wetlands Day 2004 in Crooked Tree Wildlife Sanctuary

World Wetlands Day

Olivia Carballo-Avilez

Each year on February 2nd, Wetlands Day is celebrated in commemoration of the signing of the Ramsar Convention on Wetlands. Since Crooked Tree is a Ramsar site, BAS did not lose the opportunity to take this day to teach young students of Biscayne and Lemonal villages that "wetlands are anything but wastelands." A wetland is an area that has both land and aquatic characteristics that provide rich habitat for a wide variety of plants, invertebrates and animals, including many rare and threatened or endangered species.

World Wetlands Day 2004 was celebrated in the Crooked Tree Wildlife Sanctuary (CTWS) with students and their teachers from CTWS buffer communities.

The World Wetlands Day 2006 theme is "Her future at risk, in the face of poverty … wetlands are life lines." A team including CTWS Park Director Donald Tillett, BAS Volunteer Susana Laruccia, BAS Publicity Coordinator Dirk Francisco, and BAS Education Manager Olivia Carballo-Avilez went to Lemonal Primary School and Biscayne Primary School to demonstrate to the students why the Crooked Tree Wildlife Sanctuary is at risk and the values of such an ecosystem.

World Wetlands Day 2009

BAS joined the rest of the world in celebrating the 2009 World Wetlands Day under the theme: "From Mountains to the Seas Wetlands that Work for us". BAS hosted a leaders visit to the Crooked Tree Wildlife Sanctuary. Invited guests included the Governor General his Excellency Sir Colville Young, Hon. Edmund Castro – Area Representative and Minister of State in the Ministry of Works, Hon. Gaspar Vega represented by CEO Beverly Castillo, Chief Forest Officer – Wilber Sabido, Village Councils of Crooked Tree and Lemonal Village, as well as teachers, children, and other community leaders. The Belize Audubon Society conducted this activity in an effort to bring leaders together to talk about the importance of Crooked Tree Wildlife Sanctuary and the need to plan for development.

(top) Hon. Edmund Castro addresses the 2009 World Wetlands Day audience. (bottom, from left) Hon. Edmund Castro, the Governor General Sir Dr. Colville Young, and CEO Beverly Castillo at the Wetlands Day Ceremony.

World Wetlands Day 2009

To celebrate World Wetlands Day 2009, the children from Crooked Tree Wildlife Sanctuary buffer communities Lemonal, Biscayne, and Crooked Tree went to Belize City to tell the city children about wetlands in a totally entertaining way. They played to a full house of Belize City students. A recorder group from the Crooked Tree Government School opened the show with the National Anthem (*below left*). Students from Lemonal Anglican School presented "What is a Wetland?" The wetlands campaign theme "Crooked Tree ga we baak" was dramatized by Crooked Tree Government School students. Lemonal Anglican School students presented "Protect our Wildlife" in drama and song (*above*). "Tilapia and Cow Dilemma" was presented by the Biscayne Government School students (*below right*).

Judy Lumb

CTWS Bird Festival

Olivia Carballo-Avilez

Let's protect our feathered-friends. "Sling shot? Puss Killa? or pellet gun? Which do you think will kill that Rama-tu-tu (black headed trogon) best?" This would be a typical question of children in their spare time especially during summer. Of course, why not? Belize has tons of birds—596 species.

Funded by the U.S. National Fish and Wildlife Fund, BAS held an educational Bird Festival for children of buffer communities for Crooked Tree Wildlife Sanctuary on June 10th, 2005. The 160 participating children accompanied by their teachers came from Maypen Government School, Biscayne Government School, Crooked Tree Government School and Lemonal Anglican School. The children engaged in specially designed activities under the themes: Bird Identification by Shape and Colour, Bird Identification by Bill or Beak, Bird Identification by Song or Call, Bird Conservation, Bird Habitats and Migratory Birds.

Bird Festival for children of the Crooked Tree Wildlife Sanctuary buffer communities: (*bottom*) Olivia Carballo-Avilez demonstrates identification of bird by bill or beak.

Junior Bird Club

Olivia Carballo-Avilez

In an effort to continue outreach and education beyond the Crooked Tree Wetland Campaign, a Junior Bird and Gardening Club was established through partnership with the Crooked Tree Government School. The Bird Club aims are to foster appreciation for birds, discourage killing birds for fun, and develop birding skills among children for future interest in ornithology or bird guiding. The Gardening Club aims are to get the children involved in planting local plants and vegetables that are important for birds and people. They are being encouraged to use cattle dung as fertilizer.

Through the partnership with the U.S. National Audubon Society, BAS was able to obtain a Birding Backpack for each of the 13 members of the group. The backpack includes: back pack, a pair of binoculars, birding field guide, CTWS bird checklists, T-shirts and a notebook. The National Audubon Society through the sponsorship of Allan Model have donated a spotting scope for the Crooked Tree Wildlife Sanctuary which is also being used by the birding club. Also recently donated are bird feeders for bird observation. These birding backpacks will remain in custody of the wardens at CTWS. Gardening tools were also obtained.

Four birding sessions have been conducted by the CTWS staff, Rennie Jones, Derick Hendy and Steve Tillett. They engage the children in bird games, such as bird bingo, recognizing birds from their calls, bird identification, and participation in bird discussions or presentations.

(*above*) Derick Hendy explains the bird feeder that he had in his hand to the Crooked Tree Bird Club. The group is ready to set up the bird feeder and then go on a bird walk.
(*below*) Roberto Leal, volunteer from the University of Belize shows the spotting scope donated by the National Audubon Society (U.S.).

Environmental Youth Camps

Belize Youth Conservation Corps

In the Summer of 1991, the Belize Youth Conservation Corps (BYCC) held two summer camps at the CBWS. They worked on a number of conservation projects. Several bridges were built or repaired, steps were constructed on steep hillsides, new trails were cut, and a garbage pit was dug. In 1993 twelve young ladies were at the GNP participating in BYCC-sponsored conservation activities. They built nesting boxes for birds, roosting boxes for bats, and made hanging baskets for orchids. In 1994 BYCC youth were back at GNP constructing a self-guided trail, upgrading the orchid display and repairing the steps leading down to the Belize River Recreation Centre. In June of 1995 an energetic team from BYCC cleaned the beach and cleared trails on Half Moon Caye.

Youth Enhancement Service Discovery Camp

In the Summer of 1996, 20 inner city youths along with their leaders, spent 6 days in CBWS. The objective was to use the outdoors as a practical classroom while examining and learning about the individual and group dynamics that arise as the kids worked through environmental activities together.

BAS Archives

Young Women's Christian Association's Environmental Education and Awareness Summer Camp at Crooked Tree

YWCA Camp at CTWS

About 100 youths participated in the Young Women's Christian Association's Environmental Education and Awareness Summer Camp at Crooked Tree Village. The camp, which lasted for three weeks in July of 1997, engaged the young people in meaningful environmental activities designed to help them become more involved in the sound use and management of Belize's natural resources.

Belize Council for the Visually Impaired Summer Camp

A residential summer camp for blind and visually impaired children was held at the Cockscomb Basin Wildlife Sanctuary in 1997. The camp was organized and run by BAS and the Belize Council for the Visually Impaired (BCVI) with funding from the British High Commission. The 21 blind and vision-impaired children, along with six of their brothers and sisters, got involved in a range of activities that were designed to increase independence and confidence and to develop skills for daily living, education and mobility. Activities included arts and crafts, sports and games, jungle treks and day trips. For many of the children this was their first experience of a long trip away from home.

Environmental Leadership Camp at CBWS

BAS assisted the Youth Department in a week-long Environmental Leadership Summer Camp at the Cockscomb Basin Wildlife Sanctuary in 1997. This camp, the second of its kind, was coordinated in part by youths from the previous year's group.

Fifteen youths participated, ranging in age from 14 to 17 years.

The object of the camp was to bring together a diverse group of youths from many social, cultural and economic backgrounds to spend a week camping and taking part in a wide variety of conservation-related activities.

YMCA Camp at CBWS

The Young Men's Christian Association, in collaboration with BAS, held a week-long Summer Camp at the Cockscomb Basin Wildlife Sanctuary in 1997. Twenty-four chosen participants, primarily from the Belize City area, learned abut the history of the Sanctuary, the medicinal values of plants, identification of common birds and insects, food chains and the biology of the jaguar and the black howler monkey.

Summer Day Camps

Summer Day Camps have become a staple for BAS members as every summer several innovative camps are planned for BAS members and members of the general public as well.

Sponsored by the National Fish and Wildlife Project, summer day camp was held at SHBNP on July 5th, 2005. Eighty-four children from Armenia Village between the ages of 7–12 enjoyed caving in St. Herman's Cave, swimming in the Blue Hole (guarded by a volunteer Life Guard from Jaguar Creek), bird watching, and bird identification.

A new Summer Day Camp focused on bird identification for children of BAS members was held at SHBHNP on August

(*above*) Olivia Carballo-Avilez with campers at St. Herman's Blue Hole National Park (SHBHNP) in the Blue Hole

(*below, back*) Israel Manzanero, Jr.; (*left*) Alex Escalante and Transito Jimenez, Olivia Carballo-Avilez and Israel Manzanero Sr. (*right*) with 2005 campers at SHBHNP

(*above*) Israel Manzanero, Sr. teaching campers about birds at St. Herman's Blue Hole National Park (SHBHNP)

(*below, right*) 2005 campers in St. Hermans' Cave
(*below, left*) Hand painting in Crooked Tree

18th, 2005. The highlight was a nature walk in the park and the cave by some of the best guides in Belize. BAS Board member Lascelle Tillett provided a bus.

Belize Audubon Society held two day camps in 2009. "Birds of a Feather Flock Together," was held August 9, in Belize City, led by Rudi Burgos. It was a "dollar camp" – cost only ONE Belize Dollar! From 6:30 am until 9:00 am, 32 people from the ages of 6 to 60 years did a bird walk around Belize City. "Crystal Cave Adventure Camp" was held August 22. Fifteen people spent the day enjoying the majestic experience of the beautiful Crystal Cave.

Hand Painting in Crooked Tree

Social and environmental issues were the topics of a week-long 2009 Summer Camp spear-headed by Belize Family Life Association. On July 10th BAS joined in the fun with wetland activities. The youths participated in wetland exploratory experiments, a boat trip through the wetlands, and hands as canvas art.

Educational Materials

An important aspect of BAS' environmental education has been the publication of educational materials. In the early years materials were prepared for the Radio Spots and reprinted by mimeograph. They were distributed free of charge to the schools, at exhibitions and upon request. BAS contributed fauna and flora maps for the *Atlas of Belize* produced by Cubola Press. ***Why Bother with Birds or How to Make a Bird Garden,*** a mimeographed booklet written and illustrated by Diane Hasse in October 1974, was BAS' first publication.

BAS Bulletin/Newsletter

The *BAS Bulletin* was produced monthly from 1969–1986 by Lydia Waight. It was re-designed in 1987 as the quarterly *BAS Newsletter*. Judy Lumb began as the Layout Editor of the *Newsletter* in 1989 and became Editor upon the retirement of Lydia Waight. In 2000 the Staff/Board Newsletter Committee decided that each *Newsletter* would have themes as shown in the table (*right*).

A *BAS Newsletter* insert for children called "Kid's Corner" was introduced in 2005. Educational games are developed that fit with the themes of each *Newsletter* issue. The slogan for the Kid's Corner is "Where learning about Nature is Fun!"

BAS Newsletter Themes

32:3	2000	Aug	Co-management
32:4	2001	Jan	Belize's Wildlife
33:1	2001	Apr	Renewable Energy
33:2	2001	Aug	Protected Areas
33:3	2001	Dec	Land
34:1	2002	Jun	Aquaculture
34:2	2002	Oct	Waste Management
34:3	2002	Dec	Protected Areas
35:1	2003	Jun	Marine Protected Areas
35:2	2003	July	Global Warming
35:3	2003	Oct	Crooked Tree Wildlife Sanctuary
35:4	2004	Jan	Cockscomb Basin Wildlife Sanctuary
36:1	2004	Apr	Wetlands
36:2	2004	Jul	Birds of Belize
36:3	2004	Oct	Cayo Parks
36:4	2005	Feb	Review of 2004
37:1	2005	May	Advocacy
37:2	2005	Jul	BirdLife Meeting in Belize
37:3	2005	Oct	Belize's Marine Resources
37:4	2006	Apr	Impact of Human Population
38:1	2006	Apr	Environmental Education
38:2	2006	Jul	Protected Areas and Oil
38:3	2006	Nov	Funding of Protected Areas Management
38:4	2007	Apr	Water Resources
39:1	2007	Oct	Development and the Environment
39:2	2008	Mar	Community Development
39:3	2008	Apr	Research I
39:4	2008	May	Research II
40:1	2008	Sep	Governance: Internal and External
40:2	2009	Jan	International Year of the Reef
40:3	2009	Mar	Crooked Tree Wildlife Sanctuary
40:4	2009	Apr	BAS 40th Anniversary
41:1	2009	Sep	Climate Change: Copenhagen COP-16
41:2	2010	Apr	In Honour of BAS Members

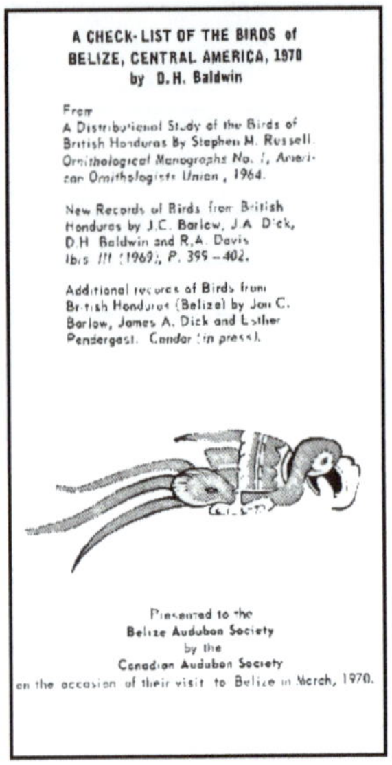

Bird Checklists

The first ***Checklist of the Birds of Belize*** was compiled by D. H. (Don) Baldwin of Toronto, Canada. He led a tour of the Canadian Audubon Society to Belize (then British Honduras) in March 1970 and presented 2,000 copies of the checklist to the Belize Audubon Society. The cover of the booklet is a picture of a Scarlet Macaw that Dr. David Pendergast of the Royal Ontario Museum discovered drawn on a pottery vase at Altun Ha.

BAS founding members Ford Young and Dora Weyer enlarged and updated the checklist in 1977. Published by Russ Mason's Flying Carpet Tours, this version had a Jabiru on the cover and recorded 67 bird families with 518 species. A few additions were made in 1983 and the checklist, which featured the Keel-billed Toucan, our national bird, was reprinted by International Expeditions. A revised checklist was written by D. Scott Wood, Robert C. Leberman, and Dora Weyer and published in 1986 by the Carnegie Museum of Natural History. Belize was divided into six geographic areas and the abundance of each bird species was given for each area.

In 1998 a new checklist was released with a Magnificent Frigatebird on the cover and 540 confirmed species. It was compiled by Bruce and Carolyn Miller and published by the Belize Audubon Society. The checklist by Bruce and Carolyn was revised and republished in 2000. Ten additional species had been verified. A "status" listing was added to indicate species of conservation concern.

In 2007, with the help of Peace Corps Volunteer Scott Schmidt, BAS produced updated bird checklists for all of the BAS-managed protected areas.

BAS Books

The Coral Reef was produced in 1986 by Janet Gibson and the Reef Preservation Committee in cooperation with the Curriculum Development Unit of the Ministry of Education and published with funds from the Reef Preservation Fund, which was started by Captain Ted Martin's "Swim for the Reef."

A to Z of Belizean Wildlife was first published by BAS under a grant for environmental education from the World Wildlife Fund International in 1986. This handbook written for students provided information and illustrations on some of the reptiles, birds and mammals commonly found in Belize. It was mimeographed by the Curriculum Development Unit of the Ministry of Education for use in the schools.

The improved second edition with a lovely colour painting on the cover was produced by BAS Environmental Education Coordinator Clara Cuellar in 1995. It was sponsored by the British High Commission and printed by BRC Printing of Benque Viejo, Belize.

The Community Baboon Sanctuary: A Guide Book was written by Robert Horwich and Jonathan Lyon and published by Orang-utan Press in 1985. The book tells the history of the Community Baboon Sanctuary and describes the black howler monkeys, along with the rainforest, flora and fauna of the Sanctuary. A second edition was published in 1987. This book was expanded and republished in 1990 under the title ***A Belizean Rainforest.***

Snakes of Belize, written by herpetologist and BAS founder Dora Weyer and illustrated by Ellen McRae, was published by the Belize Audubon Society, thanks to financial support from the World

Wildlife Fund. It was released in September of 1990. This handbook of the snakes of Belize is intended primarily as educational material for schools, but it is of interest to anyone wanting to identify the snakes found in Belize.

Tropical Forests and their Conservation: An Education Manual for Standards IV, V, VI was written and illustrated by volunteer Kerrien Redington from Cornwall, England. Kerry spent two months in the CBWS working on her book, which was published for BAS by *Producciones de la Hamaca* and printed by The Angelus Press Ltd. The manual provides an introduction to the basic science and conservation issues of tropical forests in terms that children can understand. *Tropical Forests* was first released in the Spring of 1993. The second printing in 1996 had a more colourful cover.

The Life History of the Nassau Grouper in Belize by Jacque Carter and Greg Marrow was produced in 1994 jointly by the Belize Fisheries Department, Ministry of Agriculture and Fisheries, and the Belize Audubon Society with financial support from the World Wildlife Fund. It was printed in Belize by The Angelus Press Ltd. This handbook contains helpful illustrations to explain the life of a Nassau Grouper.

Tapir Mountain Nature Reserve and Tropical Forests Conservation Education Manual was launched in 1996. It was written by Peace Corps volunteer Amy E. Smith, illustrated by Mike Mauricio and printed by BRC Printing, Benque Viejo, Belize, sponsored by The Nature Conservancy. This educational manual is full of information and illustrations on what can be found in tropical forests in general and the Tapir Mountain Nature Reserve in particular.

Cockscomb Basin Wildlife Sanctuary: A Guide to the History, Flora, and Fauna of the Cockscomb Basin Wildlife Sanctuary for Visitors, Teachers and Scientists was launched on January 29, 1997. This 350-page book was the work of 23 contributors, authors and artists, edited by Judy Lumb, and published for BAS jointly by *Producciones de la Hamaca* of Caye Caulker and Orang-utan Press of Gays Mill, Wisconsin, USA. Descriptions of the history, flora and fauna of the Cockscomb Basin are included along with maps and guides to the trails of the CBWS and an extensive environmental education section.

Coastal Treasures of Belize was launched in the same ceremony on January 29, 1997. It was prepared by Lisel Alamilla and Anna D. Hoare, funded by Coral Caye Conservation under the European Union Fund Project and printed in full colour by The Angelus Press Ltd. It is an ideal source of information for teachers and students alike, or anyone interested in learning about the marvels and significance of Belize's coastal zone and related ecosystems.

100 Birds of Belize, published in 1995, features the most common, rare and interesting birds of Belize with beautiful colour photographs, maps and text by ornithologist and photographer Carolyn M. Miller, along with 14 other artists. The birds are described in detail including their habitat, nesting habits, and other notable behaviours. Sponsors included Wildlife Conservation Society, American Bird Conservancy, Vireo, BirdLife International, Massachusetts Audubon Society, Terra Foundation, Tropical Flora and Fauna Preservation Society, U. S. Fish and Wildlife, World Wildlife Fund.

101 Birds of Belize is the second edition of *100 Birds of Belize*, also by Carolyn M. Miller. The 101st bird is the Harpy Eagle, which was reintroduced to Belize in a project spearheaded by the Belize Zoo and Tropical Education Centre. Alejandro Grajal donated

the spectacular painting of a Harpy Eagle for the cover and a number of black and white drawings to illustrate the general discussion of bird morphology and behaviour. Beautiful paintings by Fernando Zavala illustrate the 101 selected birds. This book was sponsored by National Audubon Society.

Mammals of Belize: A Checklist was compiled by Timothy J. McCarthy with illustrations by Eustorgio Mendez. It was edited by Lydia Waight and Judy Lumb, published in 1998 for BAS by *Producciones de la Hamaca*, Caye Caulker, Belize, and printed by BRC Printing, Benque Viejo, Belize. It provides a checklist with detailed illustrations useful in the identification of mammals in Belize.

Brochures

BAS' first brochure was produced in 1982. It included sections on what BAS is, the work of the Society, and why one should join. This brochure was used until 2008. Executive Director Micky Craig produced brochures for each of the parks that BAS managed and a brochure for the Reef Preservation Committee on the three species of marine turtles found in Belize.

Full colour brochures were developed in 2002 for all the BAS-managed protected areas with photographs from James Beveridge, Dirk Francisco, Anna Rossington, and Julie Robinson.

In 1990 BAS produced the brochure, ***Belize: A Guide to the Country and Its Wildlife*** with support from Programme for Belize (*left*). The guide described Belize's history, culture, geography and the major sectors of agriculture, forestry and fisheries. The brochure opened out to form a colourful poster on the Birds of Belize. New park brochures were created in 2005 and new souvenir tickets were created for Half Moon Caye and Blue Hole Natural Monuments after the entry fees were increased.

To commemorate the tenth anniversary of the declaration of the Belize Barrier Reef Reserve system as a World Heritage Site in 2006, BAS Education Programme developed exercise books (*top right*), an educational slideshow CD, and a poster.

A Community Booklet was produced in 2007 entitled "Working in Partnership with Communities for the Conservation of Protected Areas" to document BAS work with communities to garner support for the protected areas by creating opportunities for the local community to benefit from conservation practices.

In 2008 the Belize Audubon Society produced a new brochure. The Brochure features a brief description of BAS, the Protected Areas and programs managed by BAS and encourages readers to become a member (*centre right*).

BAS started the Green Initiative on Earth Day 2008 with the goal to get Belizeans involved in living green and changing a few simple behaviours that are not friendly to the environment. The brochure "Keep Belize Green by going Green: It makes ene" was developed (*bottom right*). A second brochure was produced under the Green Initiative Project for use by businesses and stakeholders in the tourism industry. This brochure promotes water and energy conservation and gives tips on how this could be achieved.

In 2009 BAS explored the idea of producing miniature Brochures of the nine BAS managed parks. The small size was seen as more cost efficient and effective in producing and also attractive to the public. A pencil/pen holder made out of paper was also developed in 2009 featuring the nine BAS managed parks.

A beautifully designed folder features the BAS logo, Mission and Vision statements as well as the 9 BAS managed parks (*top, next page*). This folder serves as a useful paper holder as well as a means of promoting BAS and the protected areas.

Colouring/Comic Books

My Reef Colouring Book were produced in 1986 by Janet Gibson and the Reef Preservation Committee in cooperation with the Curriculum Development Unit of the Ministry of Education. *My Reef Colouring Book* makes learning about the reef enjoyable for students.

In 2009 the BAS Education Programme developed three new colouring books for school children ages 5 through 12: ***Amazing Tree*** (covering biodiversity), ***Explore the Reef***, and ***Wonderful Wetlands*** (*below, left*). All are available on the BAS website.

Flight to a New Beginning is intended to raise awareness about the best practices of the cattle farming around the wetlands of Crooked Tree Wildlife Sanctuary. The techniques suggested through the comic booklet are recommendations from the Belize Agriculture Department and Belize Audubon Society. This booklet was developed by BAS with funding from National Audubon Society and the U.S. Fish and Wildlife Service.

Posters

In 1983 Rare Animal Relief Effort (RARE) President Gerald Lieberman donated posters showing a large painting of a jaguar along with a conservation message in English and Spanish.

In 1983 Esso Standard Oil Ltd. donated sets of four posters, showing a jaguar, toucan, tapir, and a yellow-headed parrot. Some of the posters were distributed free to the schools and some were sold to provide funds for reprinting.

In 1985 Dr. C. J. McCoy of the Carnegie Museum of Natural History made a donation for the printing of two turtle posters designed by BAS.

The BAS Reef Preservation Committee, in conjunction with the Caye Caulker Branch of the Belize Tourism Industry Association, produced a poster in 1987, emphasizing the need to protect the Belize Barrier Reef and its attendant lagoon ecosystems including mangroves, sea grass beds and caves.

Also in 1987, a large poster showing a drawing of the Jabiru, with information about the bird, was designed by BAS.

In 1991 a colourful reef poster was donated by *Polaris*, a cruise ship of Special Expeditions, New York. It was designed by the Half Moon Caye/Reef Preservation Committee and sold to raise money for the Reef Preservation Fund.

In 1993 BAS paid for printing a poster entitled "It All Began with the Boobies" designed by Gerald Chavannes.

A margay poster was designed by Tony Rath using one of his photographs and funded by Esso Standard Oil Ltd.

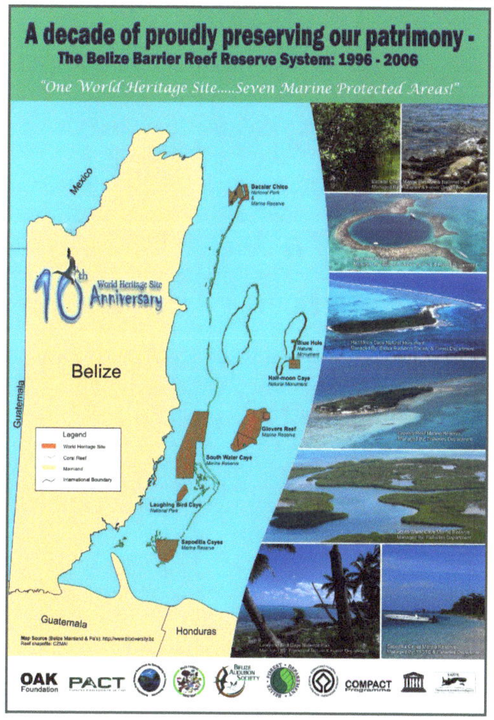

"Crooked Tree Wetlands ga wi baak; as we develop. Mek wi kip it laik dat!" This poster served as the primary campaign material for the pride wetlands campaign in 2008. It brings to the attention of readers, quick facts as to why the Crooked Tree Wildlife Sanctuary is important, while encouraging readers to be steward of the wetlands of Crooked Tree (*top, previous page*).

A jaguar poster was produced in 2008 to promote the Cockscomb Basin Wildlife Sanctuary.

A series of posters were developed to support the Blue Hole and Half Moon Caye Natural Monuments, including two for the campaign to stop the feeding of sharks at the Blue Hole (*left and below*). These posters were launched at the international Diving Equipment and Marketing Association (DEMA) Show by BAS Board Secretary Linda Searle.

BAS designed and co-produced a Fisheries Law Poster in 2009 that outlines the laws governing marine life extraction and specifically highlights extraction seasons.

BAS designed and produced a fire poster that highlights the threat posed by wildfires. Fires can destroy important forest and biodiversity. This poster was developed for the project, "Towards sustainability of BAS protected areas" with 2009 funding from PACT.

BAS also designed the primary poster for the Meso-American Society for Biology and Conservation, for the XIII congress of the Meso-American Society for Biology and Conservation.

Slide Shows and Videos

A documentary video entitled the Past, Present and the Future of fishing in the Light House Reef Atoll. This video was developed for the fishing communities of Chunox, Copper Bank and Sarteneja, all stakeholders of Half Moon Caye Natural Monument. This video highlights the history of the fishing industry in these communities, comparing the past and present situation as well as planning to ensure sustainability of the industry. This documentary was produced by BAS in 2006. This video was made possible through funding from AVINA and Oak Foundation.

A slide show was produced in 2006 to celebrate the tenth anniversary of the declaration of the Belize Barrier Reef Reserve system as a World Heritage Site. Then a seven-minute video was produced in 2009 to promote conservation at Half Moon Caye and Blue Hole Natural monument (*left*).

"Can You Imagine?" was a slideshow produced as part of the "Greening Initiative" for the 2008 Earth Day. After making its audience feel proud of our natural resources, the video goes on to present environmental problems of other countries and emphasize that Belize runs the risk of facing the same problems. If we all don't take action the future of Belize will be grim. It highlights ways in which people can get involved in making a difference: Use only what you NEED, Be ENERGY SMART, REDUCE, REUSE and RECYCLE waste, and support Belize's protected areas.

"Do you care?" was a slideshow produced by BAS in commemoration of the International Year of the Reef, 2008. It showcases marine ecosystems and how important they are for supporting biodiversity and livelihoods of local people.

NBC's Today Show broadcasting from Half Moon Caye

An educational video entitled "**Cockscomb Basin: Jaguar Proud**" was developed in 2009 by CAVU (Ceiling And Visibility Unlimited) in association with Wildlife Conservation Society, Panthera and Belize Audubon Society.

The **Brush Fire Prevention** slideshow features the importance of forests and the threat of wildfires on forest resources. It was produced by BAS as a part of the PACT funded project 2006-2009 with one activity focusing on fire management in CBWS.

A **Biodiversity** slideshow was produced for the 2009 Green Wave Campaign.

A 12-minute video entitled "**Wetlands of Crooked Tree Wildlife Sanctuary**" features a song by Leela Vernon. It was produced in 2009 by BAS as for the pride wetlands campaign.

The **BAS 40th Anniversary** video, produced in 2009 gives a history of BAS, who we are, what we do and how you can become involved

Filming at BAS-Managed Parks

BAS-managed protected areas have attracted several types of documentary producers, ranging from the educational to just letting the world see the unique features of Belize. Many films were news documentaries focussing on how BAS-managed protected areas contribute to the support of wildlife, and the alleviation of global warming.

National Geographic International (Latin America) visited Belize to document the beauty of Belize's natural resources and the impact and consequences of Belize opening its doors to cruise ship tourism. BAS' work and the challenges the Society faces in protected areas management were highlighted in the documentary, which aired across Latin America in late 2005.

Eight companies filmed in BAS-managed protected areas in 2008: Naturalight Productions, CAVU, Ushia, Azmax Filming Company, History Channel, NBC, Baltaupht Travels, and Peppermill Casinos, Inc. The highlight was when Half Moon Caye was featured on NBC's Today Show. Matt Lauer broadcasted live from the Blue Hole and Half Moon Caye Natural Monuments on November 17 and 18 as part of NBC Today's "Ends of the Earth" theme week. The Belize Audubon Society's work on turtle conservation on Half Moon Caye was highlighted, and the Marine Parks Manager Losita Lee was featured on the show. She focused on the Red-Footed Boobies on the island. Belize Audubon Society was proud to host the NBC crew at both protected areas as we know these are key conservation sites for the country of Belize and the world.

BAS on the Web

The BAS URL <belizeaudubon.org> was first registered by Nancy Tillett in 1997 and a website was created. The BAS website was expanded with a design by Anna Rossington in 2001. The option of paperless online download of the *BAS Newsletter* was added in a major update in 2006. Besides the advantage of saving trees, the web version is in full colour. BAS' website was refreshed in 2008 (*p. 119*). Members can find upcoming BAS activities in the Events Calender and participate in Advocacy or Teacher Forums to join the discussions on current environmental issues. *BAS Latest News* was added as an email Newsletter in 2008 (*right*).

Belize Audubon Society

Celebrating 40 Years

Latest News

Volume 3: Issue 4 April 2010

Biodiversity Art and Photography Competition

The Belize Audubon Society, with support from the Global Environmental Facility (GEF) Small Grant's Progam is hosting an art and photography competition. The Belizean Biodiversity Art and Photography Competition is part of a worldwide Green Wave Initiative on biodiversity. It is aimed at providing young people and Belizeans in general with useful information enabling them to participate in protecting, preserving, and improving biodiversity in their communities. The competition is open to anyone who is interested. The theme is " *Belizean Biodiversity, am I a PART of it or APART from it?"*

The competition has two categories: Artwork (which includes paintings and drawings) and Photographs.

Rules of Participation:

- Artwork or photograph must be the original work of the participant.
- Depict any life form; however it must be representative of Belizean Biodiversity.
- Entries must be a standard size of 8.5" by 11"
- Any type of art materials may be used (e.g., pens, pencils, paints including crayons).
- Only one entry per person
- On a SEPARATE sheet of paper, please write your full name and title of your entry (you can give it your own title, but it should be related to the theme stated above). In addition, please print your name at the back of the entry.

Submit entries to: Ms. Dareece Chuc, Education Program Manager Belize Audubon Society at 12 Fort Street, P.O. Box 1001, Belize City. Deadline for submissions is May 14th 2010.

Look out for these upcoming events...
✓International Biodiversity Day - May 22nd
Sponsored by the GEF Small Grants Program

BAS holds 41st AGM at Crooked Tree Wildlife Sanctuary

On Saturday, April 17, 2010 the Belize Audubon Society (BAS) held its 41st Annual General Meeting at the Bird's Eye View Lodge in the village of Crooked Tree. BAS members who attended the AGM participated in the elections of the Board of Directors. At the close of the AGM, the Board of Directors of the Belize Audubon Society for the year 2010-2011 are:

Mr. Earl Green -President
Mr. Mike Heusner -1st Vice President
Mr. Carlos Matus - 2nd Vice President

Ms. Nellita Bradley - Treasurer
Mrs. Linda Searle - 1st Secretary
Mrs. Melissa Espat - 2nd Secretary
Mr. Rudi Burgos - Director
Ms. Giselle Diaz - Director
Mr. Anicasio Garcia - Director
Mr. Mark McField - Director
Mr. Mario Teul - Director
Mr. Lascelle Tillett - Director
Mr. Carlton Young - Director

Thanks to outgoing board members Mr. Mark Hulse and Mr. Jude Lizama for their years of service. In addition to elections, presentations were made on the BAS' achievements for 2009.

The Society hosts its AGM at the protected areas it co-manages to build awareness of BAS on site management initiatives and foster appreciation for protected areas in Belize in the general public, and in particular, within BAS members. Members, who attended this year's AGM participated in activities such as birding, Cashew wine tasting and Adventure trek to Chau Hiix. Everyone in attendance enjoyed some of the special attributes that the Crooked Tree Wildlife Sanctuary has to offer. We would like to thank all who attended and contributed to making this year's AGM such a great success!

Four decades of working to maintain a balance between people and nature – join us!

For more information, please contact us at: 12 Fort Street, Belize City Telephone: 501-223-5004/4987
Website: www.belizeaudubon.org Email: membersbas@belizeaudubon.org

search belizeaudubon.org Google Custom Search Go!

Belize Audubon Society

Creating a balance between people and the environment

Home | About BAS | Protected Areas | Programs | Birds of Belize | News | Donate

"Greetings from the Belize Audubon Society (BAS) the oldest conservation organization in Belize! Did you know that the oldest protected area in Belize and in Central America is Half Moon Caye Natural Monument and that BAS manages it along with eight other key conservation areas? We invite you to explore the nine protected areas the way nature designed them and to help us support our programs".

-Executive Director Anna D. Hoare

BAS Protected Areas

The nine protected areas we care for, represent the richest land and marine environments in Belize, with critical habitat for jaguars, a wetland of international importance, impressive coral reefs, and sacred caves with beautiful limestone formations and remains of the ancient Maya.

Bird Watch

Are you interested in bird watching? If so, you don't want to miss the BAS Bird Watch. We report the latest sightings gathered from around Belize. Or learn about the our member's favorite birding hotspots in Belize.

Creating a Balance

The Belize Audubon Society is dedicated to achieving and maintaining a balance between people and the environment. We work to fulfill our mission through environmental education, advocacy, community development, research, and biodiversity conservation.

Featured Event

PAUSE II

You've seen all photos in the set **Dancing and Dining 2008**. Would you like to start again from the beginning?

NO YES

Enjoy pictures of the first ever Dancing and Dining with Nature Gala.

Pictures by: Dirk Francisco

Join Belize Audubon Society Today!

Become a Fan!

facebook

Financial Report

2006-2007 2007-2008

Auditors' Report 2008

(*top left*) Israel Manzanero, Jr. and Olivia Carballo-Avilez with 2005 Summer Camp at St. Herman's Blue Hole National Park; (*top right*) World Wetlands Day 2003 at Crooked Tree Wildlife Sanctuary; (*bottom*) 2009 World Wetlands Day at the Bliss Centre for the Performing Arts

2009 World Wetlands Day at the Bliss Performing Arts Center

Summary of BAS Environmental Education

From the beginning the founders of BAS understood the importance of environmental education, both for school children and the general public. Using the technology available in the 1970s, they prepared slide shows for schools and radio programs. Lydia Waight produced the *BAS Bulletin* every month for seventeen years with her typewriter and mimeograph machine, copies of which were delivered to all the secondary schools.

By the early 1990s funds were available for Staff, which greatly expanded the BAS Environmental Education Programme. Countless presentations have been made to schools, organizations, and special audiences. Displays have been created for many public events. Publications such as brochures, posters, books, and hands-on kits have been produced. Videos, CDs, and the BAS website spread the conservation message.

Each year BAS celebrates Earth Day, World Environment Day, World Wetlands Day, International Biodiversity Day, and other major events. The BAS Annual General Meeting provides an opportunity for education of our members and friends.

The BAS Environmental Education Programme focuses on educational activities with the buffer communities and stakeholders of the BAS-managed protected areas. An assessment of the educational landscape in Belize clearly demonstrates that BAS is the foremost environmental educator in the country.

Armenia Village youth have been trained as "the Guardians of St. Herman's Blue Hole and the Environment"; Crooked Tree children have formed the Crooked Tree Junior Bird Club; special workshops have been held for teachers in buffer and stakeholder communities; Chunox, Copper Bank, and Sarteneja fishermen have been trained in scuba diving, tour guiding, and sport fish guiding as alternate livelihoods; and community members have been trained to produce handicrafts that are sold in BAS gift shops.

These are but a few of the many contributions the BAS Environmental Education Programme makes in an effort to ensure a sustainable future for Belize.

Belize Audubon Society

Celebrating 40 Years

Chapter Four

Protected Areas

Management of Protected Areas

Throughout our 40-year history, BAS has had an important role in the management of protected areas. BAS' first project in 1969 evolved from a concern for decimation of booby chicks by introduced rats on Half Moon Caye. As our requests for particular reserves were granted, our involvement in the management was expected.

This was first documented in 1984 when BAS was given the authority *"to work along with the Forest Department of the Government of Belize in the protection and management of areas designated under the Forestry Act and the National Parks and Refuges Acts"* in a letter from James V. Hyde, Permanent Secretary, Ministry of Natural Resources.

BAS signed a five-year agreement, which defined the responsibilities for management of six protected areas, Half Moon Caye Natural Monument, Crooked Tree and Cockscomb Basin Wildlife Sanctuaries, Guanacaste and Blue Hole National Parks and Tapir Mountain Nature Reserve.

An updated agreement between BAS and the GOB was signed on February 6, 1999, which added the Blue Hole and Victoria Peak Natural Monuments.

On August 31st, 2004, a ten-year agreement was signed, which allows BAS to complete long-range planning for protected areas management. This agreement included Actun Tunichil Muknal Natural Monument, which is managed cooperatively with the Archaeology and Forest Departments.

Half Moon Caye Natural Monument

In 1820 the first lighthouse was built on Half Moon Caye at Lighthouse Reef, an atoll 60 miles east of Belize City. The caye is divided into two very distinct ecosystems, a coconut grove on the eastern end and a littoral forest on the western end. The 45-acre caye supports a nesting colony of Red-footed Boobies. The boobies co-exist with their pirate neighbours, the Magnificent Frigatebirds.

On the first of September, 1928, the western end of Half Moon Caye was made a Crown Reserve (*BAS:30 Years,* p. 121). In 1971 Ford Young, a real estate agent and BAS Vice President, was asked to sell five lots on Half Moon Caye. Mr. Young brought this information to the BAS Board, who requested that Government acquire the remaining private lots adjoining the Reserve on Half Moon Caye.

Dr. Craig MacFarland of *Centro Agronomico Tropical de Investigacion y Enseñanza* (CATIE) raised half of the funds needed to buy the lots and Government contributed the other half and cancelled several leases.

Half Moon Caye Natural Monument (HMCNM) was established as Belize's first protected area under the National Parks System Act, when Statutory Instrument No. 30 of 1982 was signed the 4th day of March, 1982.

Maintenance of HMCNM

BAS' first project was the poisoning of rats on Half Moon Caye. In August, 1969, the Society wrote a letter to the Chief Game Warden Dean Lindo expressing its concern that the nesting colony of

Funds were given by CATIE for the management of the Natural Monument, US$1500 in 1979 and over US$4,000 in 1984. Under the direction of BAS Half Moon Caye/Reef Preservation Committee, trails were cleared, an observation tower was constructed in the bird rookery, and picnic tables and a barbecue grill were built for the visitor area in the coconut grove.

David Craig got an old water tower railing from the Belize Water Authority and a ladder from the Belize Brewing Company. With these and other donated materials, he and Lascelle Arnold built the first observation tower. Four large signs showing maps of the Caye and several other signs were donated by the New York Zoological Society.

For several years maintenance on Half Moon Caye was accomplished periodically by groups of volunteers from BAS and the Fisheries Department.

Judy Lumb

Red-footed Boobies was being decimated by rats that had been introduced to the island from ships. They proposed to use the rat poison Wayfarin and asked for the Chief Game Warden's help in eliciting the cooperation of the Custom's Department. Having gained the required permissions and materials, BAS members Meg Craig, Jim Parker, and Dora Weyer, accompanied by Customs Comptroller J. J. Robateau, went to Half Moon Caye on the *Lolette* on October 8, 1969, to begin the project.

BAS Archives

Meg Craig (second from left), Dora Weyer (second from right), J. J. Robateau (right) with Mr. Garbutt, the lighthouse keeper and his children on Half Moon Caye October 9, 1969

Then in 1997 full time BAS Staff members were installed at Half Moon Caye paid by a grant from the Coastal Zone Management Project. Later two wardens were added, paid from entrance fee income. That income was significantly increased in 2005 when the fee structure was amended to increase the fees for non-Belizean visitors to Half Moon Caye and Blue Hole Natural Monuments.

The 2004 World Environment Day theme was "Wanted Seas and Ocean, Dead of Alive," so thirty-five BAS Staff and Board members and volunteers spent three days on Half Moon Caye in a major renovation project. A new radiating well system was installed and volunteers from Belize Electricity Ltd assessed the electrical supply.

Volunteers Abril Esquivel of Mexico and Gabor Vereczi of Hungary created an exhibit in the Visitor Centre, which was remodelled by Bill Borella, an artist in residence from New York (*right, top*).

World Heritage Site

The United Nations World Heritage Committee formally adopted seven protected areas along the Belize Barrier Reef and its adjacent atolls as a World Heritage Site under UNESCO at their meeting in Merida, Mexico, on December 4, 1996. HMCNM is included as part of this prestigious World Heritage Site.

HMCNM Management

The first management plan for the Half Moon Caye Natural Monument, was based on the draft prepared by David Lovell, a consultant from CATIE. It was presented to Government in May of 1985.

(*above*) BAS staff patrolling

(*top left*) Visitor Centre display at Half Moon Caye

(*centre left*) Kayaking at Half Moon Caye

(*bottom left*) BAS staff anchoring demarcation buoys

(below) Diving at Half Moon Caye Wall

Julianne Robinson

Julianne Robinson

(*above*) New signs on Half Moon Caye

(*below*) New pier on Half Moon Caye

The next management plan for HMCNM was written in 1995 by Diana Williams, the BAS Staff, and the Protected Areas Technical Committee. In 2006 a new management plan was completed through AVINA and OAK funding. In 2008-2009, a Public Use plan was completed through funding from TNC and UNESCO.

The "Conservation of the Lighthouse Reef Atoll" programme funded by the Summit Foundation from 2000–2003 aimed to conserve the biodiversity of the Atoll by sustaining tourism, fishing, research, and education. Moorings were installed; zones were established; and a local advisory committee was formed with representatives from stakeholders and government agencies. An underwater interpretive dive trail was developed.

A major upgrade in 2007 included a new pier and bathroom facilities funded by the Protected Areas Conservation Trust (PACT). BAS then created and installed attractive interpretive signs.

Half Moon Caye has been a favourite site for film crews. BBC was there in April of 2006 and NBC broadcast the Today Show live from Half Moon Caye in 2008.

The "Half Moon" of Half Moon Caye from the Air and the Ground

Conch Bust

Roberto Pott

Under the National Park System Act 1981, it is illegal to fish within any natural monument. The boundaries of the Half Moon Caye Natural Monument are clearly demarcated by mooring buoys and BAS has undertaken extensive education campaigns to inform fishers of the protected areas boundaries as well as the importance of protected areas to maintaining Belize's fisheries.

The BAS marine protected areas staff detained three persons for illegally fishing within the Half Moon Caye Natural Monument. One hundred and fifty-two (152) conchs that had already been removed from the shell were confiscated for evidence, and 194 live conchs were released within the Half Moon Caye Natural Monument. The three men appeared in the Magistrate's Court and were found guilty of catching fish within a natural monument without written authorization and illegal entry into a natural monument. All three were charged $400 for the offences. The Forest Department under whose jurisdiction natural monuments fall, prosecuted the case. The confiscated conchs were donated to the Dorothy Menzies Childcare Centre in Belize City.

(*above*) Confiscated illegal conch. *Photos by Roberto Pott*

(*left*) George Hanson, Enforcement Officer with the Forest Department hands over confiscated 152 shelled conch to the Dorothy Menzies Childcare Centre in Belize City.

Blue Hole Natural Monument

At the opening ceremony of the BAS 30th anniversary Hon. John Briceño signed the Statuatory Instrument that placed the Blue Hole Natural Monument under BAS co-management.

Snorkeling the Blue Hole

(*top and centre*) Public Resource and Use workshops for the World Heritage Sites
(*bottom*) Khaida Mossiah presenting green bag to potential clients at BETEX (*photo by Marci Geban*)

Promoting Conservation through Sustainable Tourism Development

Kaidia Mossiah, Marceluz Geban, and Losita Lee

The designation of seven areas in the Belize Barrier Reef Reserve System (BBRRS) as a World Heritage Site makes it a site of international natural and cultural heritage, and a source of national pride. BAS engaged in a highly participatory programme entitled "Promoting Conservation through Sustainable Local Tourism Development", funded by The Nature Conservancy and UNESCO to plan for sustainable tourism at the Half Moon Caye and Blue Hole Natural Monuments. A series of six workshops were held in 2008 and 2009 to gather ideas and opinions from tour operators, researchers, past inhabitants of the island, visitors to the area, Half Moon Caye staff, tour guides, the Forest Department, and Tourism lecturers from the University of Belize, leading to the development of a Public Resource and Use Plan.

BAS participated in the 2009 BETEX tradeshow in San Pedro and the Diving Equipment and Marketing Association show in Las Vegas Nevada to promote conservation at the Blue Hole and Half Moon Caye through a message to Conserve, Explore, Enjoy! The booth was shared with Sea Sport Belize Scuba Diving, a Tour Operator who frequently visits the Blue Hole and Half Moon Caye Natural Monument.

Several materials such as posters, brochures, pens, promotional CDs, directional and interpretive signs were developed to increase awareness of the importance of Blue Hole and Half Moon Caye and distributed at both tradeshows. A "Limits of Acceptable Change" study was carried out at Half Moon Caye and Blue Hole Natural Monuments particularly looking at the status of the coral reef communities (*Chapter Five*).

Public Resources and Use Plan Launched

On July 15, 2009, the Public Resource and Use Plan for Half Moon Caye and Blue Hole Natural Monuments was launched. Many are attracted to these sites, intrigued by their designation and interested in experiencing their outstanding value. Consequently there is an increase in activity in and around the sites as visitors want to experience all the sites have to offer. In light of this BAS, the Forest Department, stakeholders, and partners, have been engaged in a two-year public use planning process. Interested persons provided input, examined visitor experiences, validated the site goals and objectives, identified tourism attractions, selected the "best" products, and fine-tuned regulations.

(*above*) BAS Executive Director Anna D. Hoare presents the Public Resource and Use Plan to Protected Areas Officer, Raul Chun, who represented the Forest Department.

The Plan facilitates, promotes and plans for sustainable tourism in protected areas based on responsible management for conservation. It serves as a tool to guide conservation of the two sites, encouraging visitors to be engaged in their experience while ultimately ensuring the integrity of the area.

The onus is on BAS, the management agency, to implement the Plan and control user behaviour to ensure that the integrity of the area is maintained.

The project included the development of supporting materials. *Best Practices Booklet* guides the stakeholders in the sustainable use of the two Natural Monuments. A DVD was created that dramatically demonstrates the value and importance of the Half Moon Caye and the Blue Hole. An attractive poster entitled "Protect the Fishing Industry and Marine Life of Belize" describes the biodiversity of the marine life in the World Heritage sites.

Judy Lumb

Boat-billed Heron (*Cochlearius cochlearius*)

Crown Reserve Bird Sanctuaries

In 1975 the Government of Belize reserved Bird Caye in Northern Lagoon Manatee (not to be confused with the Northern Lagoon which is at Crooked Tree) as a bird sanctuary. In April of that year a BAS field trip group put up signs painted by Board member Shelly Matus alerting the public to the Reserve status of the caye and warning against hunting, shooting, and landing on the Caye. After the signs had washed away, they were replaced by Bert Forman in October of 1979.

The BAS continued their vigilance over this Reserve and lobbied for several years to get protection for other bird rookeries in Belize. Finally, in the Gazette of September 17 1977 (No. 39), seven cayes, including Bird Caye, which were rookeries for wood storks; great, cattle, and reddish egrets; boat-billed and tri-coloured herons; white ibis; frigatebirds; anhingas and other birds were declared as reserves.

Lionel (Chocolate) Heredia was taking tourists from Caye Caulker to Gales Point and stopped at the Bird Caye to show his guests the birds. He was concerned when he saw hunters At that time BAS had no funds to pay wardens, but Chocolate was not asking to be paid. He just wanted the authority to interrupt any hunting activities that he encountered. So BAS appointed him as an Honorary Warden of Bird Caye, Northern Lagoon, Manatee on June 21, 1982.

Columbia Forest Wildlife Refuge

Dora Weyer had the idea of reserving part of the Columbia Forest in the Toledo District as a wildlife refuge. The Florida Audubon Society and Florida State University wanted to use it for research and education. The Government agreed to declare 6,000 acres as a wildlife refuge and the area was demarcated in 1972 with BAS members donating the labour, fuel and oil. But the wildlife refuge was never declared. A rice cultivation project in that area took precedence.

Bladen Branch Nature Reserve

Bladen Branch Nature Reserve is a 97,000-acre reserve is situated in the Toledo District along the Main Divide of the Maya Mountains. Although the BAS never had official management responsibility for the Bladen Branch Nature Reserve, it was assumed and BAS did act in that capacity for some years. A Rapid Ecological Assessment of the area was carried out in March of 1993 by students and staff members from the BAS and The Nature Conservancy.

Guanacaste National Park

It was through the good offices of founding member Magnus Halcrow that the Society's request for reserving the 55-acre parcel of land at the junction of the Western Highway and Roaring Creek was granted. A dentist, Dr. Otto Faulkner, had applied for this property, but when he learned that it was to be reserved, he generously sent a contribution to BAS towards its development. The Society applied to Government asking to purchase it, but Mr. Halcrow did some lobbying. BAS did not have to buy the land because in 1973 they were given a Conditional Free Grant for Guanacaste Park on condition that is be used only for a bird sanctuary or national park.

The area is located at the confluence of the Belize River and Roaring Creek, and the Western Highway. With Government approval, was named Guanacaste Park for the very large, old Guanacaste or tubroos tree (*Enterolobium cyclocarpum*) growing there. This large tree supported a variety of epiphytes. The park also has numerous species of trees and plants, all of which are preserved in their natural state.

The famous guanacaste tree that gives GNP its name.

For many years, Guanacaste Park was managed by volunteer labour and donated materials. In 1974, Goshen College students cleared the first paths and made fences. Participants on a birding outing erected warning signs. In 1978 Hurricane Greta did considerable damage in Guanacaste Park, so clean-up required a big effort by BAS volunteer work groups.

After 1987 there were BAS personnel at GNP full time, beginning with Peace Corps Volunteer Matthew Miller who planned World Environment Day ceremonies at Guanacaste Park with Hon. Dean Lindo, Minister of Agriculture, as the main speaker.

The next year a MacArthur Foundation grant allowed BAS to employ wardens for the Guanacaste Park and the Blue Hole National Park. Martin Ack, Juan Choc and Heraldo Ack, were the first Cayo Park wardens. Peace Corps volunteer Robert Habeck worked to help develop Guanacaste Park and Blue Hole National Park.

Welcome to GNP

Guanacaste Park became Guanacaste National Park (GNP) on the World Environment Day April 22, 1990, when Hon. Florencio Marin, Minister of Industry and Natural Resources signed the Statutory Instrument that declared the area a National Park, which cancelled the Conditional Free Grant.

A preliminary guide to the trees in the park was prepared in 1991 by Matthew Miller and illustrated by Laura Callahan. A list of birds was compiled by Bruce and Carolyn Miller and Chandler Robbins. A beautiful butterfly display was prepared and donated by Jan Meerman, Manager of Shipstern Nature Reserve.

Villagers from nearby Roaring Creek participated in a Community Workshop in GNP in October of 1992. They formed the "Friends of Guanacaste" support group and provided valuable input into the Management Plan for the GNP, which was completed in 1993 by a committee headed by Peace Corps Volunteer Robert Mackler.

Visitor Centre at GNP

Eugenio Ah, Head Warden of the Cayo Parks, organized an Earth Day poster contest in 1993, "Why is a Tree Important?" for the Roaring Creek School. Remarkable talent was exhibited in the artwork.

The educational display in the Guanacaste Visitor Centre was prepared by Peace Corps volunteers Bonnie Gestring and Rebecca Nealey. It was officially opened on March 18, 1994, and dedicated to Peace Corps Service in Belize in recognition of their assistance to BAS. An interpretive booklet for the nature trail at GNP was developed by Michael Somerville, BAS Environmental Education Coordinator, assisted by Steven Creamer, a volunteer working in BAS office during the Spring of 1997.

GNP was the destination of the Walkathon and site for the Opening of the 30th Anniversary Celebrations on February 6, 1999.

Owing to its central location, GNP has been a favourite location for school visits. Thousands of children enjoy the educational experience of GNP each year (*above*).

New Observation/Swimming Deck

In 2000 the flooding of Roaring Creek from Hurricane Keith totally destroyed the existing swimming deck in GNP. The new observation/ swimming deck with steps leading to the river for easy access to the creek was built by Nexworld Industries and funded by PACT (*right*).

The deck was scheduled to be officially opened on the of 28th June 2002. However, due to major flooding, the event had to be cancelled. But the deck withstood the test of one of Belize's most destructive flash floods where it was under water for close to three weeks. In 2005, a new bird deck funded by U.S. National Fish and Wildlife Foundation was constructed using the existing bridge foundation to provide a classroom setting for visiting classes.

Guanacaste National Park as Nature School

Olivia Carballo-Avilez

37th AGM at Guanacaste National Park

Judy Lumb

Just before the 37th AGM at GNP in 2006, the Guanacaste tree, the signature tree for the park, had to be cut for safety reasons, particularly since thousands of school children visit the park. One of the three large branches had fallen five years ago and there was concern about the remaining two, so BAS consulted Forest Department experts to determine their status. When it became clear that both were quite rotten, the difficult decision was made to cut them before one fell and injured someone. The 20-foot high trunk is now all that is left of that once grandiose tree (*left*). The AGM at GNP gave BAS members the opportunity to see the remains of the famous tree. The good news is that there is another huge Guanacaste tree on the River Path that will become the new signature tree in Guanacaste National Park.

St. Herman's Blue Hole National Park (formerly Blue Hole National Park)

Lydia Waight

The Blue Hole on the Hummingbird Highway was a favorite destination for field trips organized by Vice President Ford Young. In addition to the opportunity for a refreshing dip in the pool, there was good birding. The Blue Hole is a sinkhole in the limestone and has a small stream emerging from a siphon on one side and disappearing into a cave on the other side. In 1972 we were pleasantly surprised to find that neat concrete steps with a good, solid handrail had been built from the highway down to the bottom of the sink hole. We were also very pleased to note that the surrounding vegetation had not been destroyed. Ten years later the steps needed attention again, so the Society requested that the steps be replaced and the area be protected.

The 575-acre Blue Hole National Park (BHNP) was established when S.I. No. 109 of 1986 was signed by Hon. Dean Lindo, Minister of Agriculture, on November 23, 1986. BHNP has been managed by the Belize Audubon Society ever since and includes the Blue Hole, a sapphire-coloured natural swimming spring, and St. Herman's Cave. A landscape artist from Panama sponsored by CATIE assisted in the design of use areas and a trail system for BHNP.

In 1988 funds from the MacArthur Foundation were used to construct a retaining wall to prevent erosion around the concrete stairway. A trail was cleared from the Blue Hole to the opening of St. Herman's Cave. The grant also provided salaries for wardens for both the GNP and the BHNP. A committee headed by Peace Corps Volunteer Robert Mackler prepared the first Management Plan for BHNP, completed in 1994.

Raleigh Volunteers in BHNP

In 1999 Raleigh Volunteers spent three weeks in the BHNP opening the northern boundary of the park and relaxing in the Blue Hole after work each day.

Adjacent Private Reserve

An agreement was signed on November 15, 1994, between the Eden Conservancy, through their Target Earth programme, and the Belize Audubon Society to permanently conserve land adjacent to the Blue Hole National Park. Eden raised the funds to purchase 3,400 acres, which they conveyed to BAS in 2003. BAS manages the property within the management plan developed for BHNP.

Name Change to St. Herman's Blue Hole National Park

Tourists were often confused between the Blue Hole Natural Monument in Lighthouse Reef and the inland Blue Hole National Park. To solve this problem, the name of the Blue Hole National Park to the St. Herman's Blue Hole National Park in 2004.

SHBHNP Visitor Center

The Visitor Centre at SHBHNP was upgraded with displays created by Bill Borella with assistance from two Gap Challenge volunteers, Katharine Davis and Siobhan Cox. The main features are the artist's renditions of the cave systems, the Blue Hole, flora and fauna.

Birding with Warden Israel Manzanero, Sr. in SHBHNP.

SHBHNP Visitor Centre display created by Bill Borella

(*above*) Warden Israel Mazanero, Sr. conducting a tour for Ambassadors to Belize from several countries.

(*below*) Exploring St. Herman's Cave in 2002 are Minister of Natural Resources Hon. John Briceño and Director of Forestry Nigeli Sosa (*second and third from left*), along with Warden Israel Mazanero (*left*) and Cayo Parks Protected Areas Manager June Neal (*right*)

Valdemar Andrade

RARE Trail at BHNP

On November 27, 1997, the Belize Audubon Society officially inaugurated the uniquely designed RARE Trail, named for Rare Animal Relief Effort, the funding agency that made its construction possible. The trail extends for some two miles over rugged landscapes, and through the lush tropical forests. The RARE Trail features several campsites, an array of beautifully painted interpretive signs a spectacular observation platform and majestic mahogany trees tower above the path.

Camping Trail

For guaranteed wildlife sighting, an excellent trail to walk is the Camping Trail. The camping trail is about 1.5 miles in length and of moderate difficulty to hike. Many sightings of peccary herds and howler monkey troops have been reported by both visitors and wardens along this trail.

Dusky Antbird Trail

In 2003 the Dusky Antbird Trail replaced the original trail that connected the Blue Hole with the Visitor Center and St. Herman's Cave with the assistance of Gap Challenge volunteers. Some birds sighted along this trail include the Dusky Antbird, Black-headed Trogan, Golden-crowned Warbler, Roadside Hawk, Lineated Woodpecker, Social Flycatcher, Olive-backed Euphonia and many others. Other wildlife sighted along this trail include squirrels, gibnut, peccary, fer-de-lance, and deer.

Crystal Cave

SHBHNP has another cave, the Crystal Cave or Mountain Cow Cave which goes for miles with huge caverns that are adorned with spelotem and crystalline formations. Exploring the Crystal Cave is only for the adenturous and hardy visitor. Most of the trail is uphill and upon reaching the entrance of the cave, visitors are required to rappel about 70 to 75 feet to the entrance of the cave and another 20-foot rappel into the cave. This trail is not marked since visitors to the parks are not allowed to use it without a guide.

Visitors then take a walk down the foot path and staircase of the Mayan shamen and priests as they journeyed into the underworld of *xibalba* (a place of fright), where they used chambers deep within the cave for their ceremonies. Here visitors can observe and photograph artifacts of potteries, beads, obsidian knives and human skeletons that have been calcified into the limestone floor and date back 2,000 years.

SHBHNP Business Plan

Sponsored by the Protected Areas Conservation Trust (PACT), a business plan was constructed for St. Herman's Blue Hole National Park by consultants Greg Findley and Scott Edwards of the Center for Park Management.

The total Park income in 2004 was approximately BZ $156,000 of which $45,000 were from grants. Gap Analysis indicated that $215,000 would be required for the Mission Critical level of management and $250,000 for the Optimal level. Various strategies were suggested make up the gap between the current resources and the ideal levels.

Crystal Cave formations and Mayan artifacts

St. Herman's Blue Hole National Park as Nature School

38th AGM at St. Herman's Blue Hole National Park

St. Herman's Blue Hole National Park

Tapir Mountain Nature Reserve
(formerly Society Hall Nature Reserve)

Staff house at TMNR

Striped Basilisk (*Basiliscus vittatus*)

Mrs. Svea Dietrich-Ward and her husband came to Belize in 1969. Mrs. Dietrich-Ward was looking for an area that was unspoiled by mankind so she could repay Nature for what her presence used of Earth's resources. They hired BAS Vice President Ford Young as their real estate agent. By 1974 they had found a 7,000-acre area between Barton Creek and Roaring Creek.

Most of the land (6,741 acres) was donated to the Government of Belize and declared as Society Hall Nature Reserve in 1986. Mrs. Dietrich-Ward came to the BAS Board Directors meeting on February 12, 1987, to ask for help in managing the Nature Reserve. The Government leased the land back to the Wards and that lease was transferred to BAS in August of 1990. The name was changed to Tapir Mountain Nature Reserve (TMNR) on July 16, 1994.

Peace Corps volunteer Amy Smith focussed her work on the TMNR from 1994 to 1996. She wrote a guidebook which was illustrated by Mike Maurico and published by BAS. Support was obtained from The Nature Conservancy to allow David Tzul to be hired as a warden.

International Conservation Relief organized Hike and Bike for the Rainforest, which was sponsored by Coca Cola in 1994, to raise money for TMNR. A motorcycle was also donated to TMNR.

The 1994 biological survey of TMNR recorded 129 species of migratory and resident birds, ten mammals, 160 species of plants, 21 species of damselflies and dragon-flies, 96 species of butterflies and moths, three species of fish, one amphibian species, and 12 species of reptiles.

The Friends of Tapir Mountain Programme was organized at a TMNR Community Workshop and Interpretative Field Trip on September 3, 1995, at Pook's Hill Lodge near the northeast corner of the Reserve. Seventeen village leaders, school teachers, students, and community health workers from seven nearby villages participated.

A Community Resource Center was built for Tapir Mountain Nature Reserve by Raleigh International with support from the Canada Fund Project in Belize. The Center was opened on May 5, 1997, in the presence of representatives of the buffer zone communities, Blackman Eddy, Roaring River, Ontario, Unitedville, Georgeville, Teakettle, San Antonio and Upper and Lower Barton Creek.

Damselflies are on the list of species found in TMNR.

Illegal Logging in TMNR

The first incident of illegal logging in Tapir Mountain Nature Reserve was documented in March 2004. In this instance no one was caught but signs were placed in the area and a fence erected across the access road that had been constructed.

On June 24th 2004 one BAS Staff and a casual worker encountered three men clearing an area within the boundary of the Reserve. When approached by the BAS Staff, two of the three men ran into the bushes. The third trespasser, known to the BAS Staff, tried to bribe them, but was told to leave the reserve immediately. BAS Staff found found a large mahogany tree that was being cut into lumber, as well as the horse and cart that was being used to transport the lumber out of the Reserve.

BAS staff returned the next day accompanied by two Police Officers and a Forest Officer. The trespasser was taken into custody and charged with illegal entry into a reserve and illegal possession of forest products and 15 pieces of mahogany lumber valued at $15,000 were confiscated.

Illegal logging in TMNR

Actun Tunichil Muknal Natural Monument

June Neal

The hilly karst topography TMNR is covered with subtropical moist forest, which is largely intact. The Reserve also contains sinkholes, ancient Mayan sites and Cave systems, one of which is the famous Actun Tunichil Muknal Cave.

Although found well within the boundaries of the Reserve, this particular cave was still under the management of the Department of Archeology (now known as the Institute of Archeology) and under this management two tour agencies, Paxz Tours and Maya Walk Tours were granted permission to conduct tourism activities within this particular cave.

This has led to a series of meetings and discussions between BAS and the Archeology Department. In January of 2003 it was decided that 455 acres of the Tapir Mountain Nature Reserve would be dereserved and redesignated as a "Natural Monument." Statutory Instrument (SI) # 16 of 24 of 10 January 2004 was passed, establishing this area as a natural monument and an additional SI was passed in June of 2004 that established a fee structure of $30 BZ per person. It was originally declared under the name "Tapir Mountain Natural Monument," but soon changed to "Actun Tunichil Muknal Natural Monument" (ATMNM).

It was necessary to purchase 5 acres of land from private land owner David Hayles for the construction of a guard post and for right-of-way access to control the entrance road to ATMNM.

The entrance to Actun Tunichil Muknal Cave

Official signing of Memorandum of Understanding between the Belize Audubon Society and the Institute of Archeology (IoA) occurred on June 10th 2004 at the BAS office, the first of its kind to be signed by the Society for the co-management of an archeological site.

In the MOU between BAS and IoA, both parties have agreed to jointly manage and develop the Natural Monument in accordance with the National Institute of Culture and History (NICH) and the National Park System Act and to put in place such other regulations by SI or otherwise, as may be necessary in order to ensure full implementation of the agreement.

Activities under this agreement include the development of management and operational plans, infrastructure development, fee collection, trail development, tour guide trainings and other day-to-day management of the Natural Monument. All management costs and entrance fees are divided 50/50 between BAS and IoA. Tickets to enter the reserve with these agencies are sold from the Cahal Pech Visitor Center in San Ignacio.

Every year the cave must be closed when the road and entrance to the cave are flooded. In 2006 some illegal mining resulted in the re-routing of the river, which increased flooding. The isolation of the ATM area brings challenges. Tourists have been robbed of their valuables, but a quick response resulted in the bandits being caught. Illegal xaté harvesting is also a problem. In 2007 a joint patrol with the police and Forest Department resulted in the arrest of eight xatéros.

Mayan pots (*top*) and skeleton (*bottom*) in the ATM Cave.

Judy Lumb

Crooked Tree Wildlife Sanctuary

Establishment of the Sanctuary

Lydia Waight

Mrs. Dora Weyer had visited Crooked Tree Lagoon for many years because it was always a place that had a lot of birds. In March 1970 BAS made a river trip to Crooked Tree Lagoon along with a visiting tour of the Canadian Audubon Society. The tour participants drove 16 miles up the northern highway to Double Run, where the water treatment plant is located, and boarded a flat-bottomed tunnel boat that took us to Crooked Tree Lagoon. The next year a similar trip was planned with a group from Florida.

In 1972 Alexander Sprunt IV, Head of the U.S. National Audubon Society Field Office, came to assess Crooked Tree and make recommendations about its creation as a wading bird reserve. He proposed the establishment of a Natural Area Reserve at Crooked Tree Lagoon. Copies were sent to the Premier, the Minister of Trade and Industry, who was responsible for national parks, and the Chief Forest Officer who was also the Chief Game Warden.

The first wildlife sanctuary declared by the Government of Belize, Crooked Tree Wildlife Sanctuary was gazetted on 8 December, 1984, (S.I. No. 95). This 22,000-hectare wetland area encompasses eight lagoons and Black Creek, which drains into the Belize River. The sanctuary is a subtropical moist zone and includes open water areas, lagoons and ponds with typical floating plants and high biodiversity, especially birds and aquatic communities. Since 1984 Crooked Tree Wildlife Sanctuary has been supported by the Wild Wings Foundation, which continues to provide support for its maintenance.

The Sanctuary is a critical habitat for water birds and other migratory and resident wildlife dependent on water, especially during the breeding season and during the dry season. There is considerable historic value to the area and its logwood forests, and the Sanctuary protects a gene pool for the logwood tree (*Haematoxylum campechianum*). Large bullet trees (*Bucida buceras*) grace the edges of the lagoon (*right*).

Causeway

From the time the Ministry of Works constructed a causeway across the Crooked Tree Lagoon connecting the village with the Northern Highway in 1983, BAS expressed concern that the natural flow of water through the lagoon had been interrupted. Some villagers were requesting that culverts be installed to allow free flow of the water in the lagoon. In the Fall of 1992 the problem was solved with two 20-foot-wide cuts in the causeway, allowing the natural flow of water to return to Crooked Tree Lagoon. Bridges were constructed so traffic could pass (*aerial view right*).

In 2008 another controversial issue resulted with a second causeway that was constructed, this time in the Western Lagoon (*Chapter Two*).

(*top*) Crooked Tree Lagoon and bullet tree
(*bottom*) Aerial view of Crooked Tree causeway

(*top*) CTWS Warden Rennie Jones leads children on a bird walk.
(*bottom*) CTWS Visitor Centre

Management of the Sanctuary

In February of 1985 the first tentative operational plan for the Crooked Tree Wildlife Sanctuary was drafted by David Lovell. From April to June of 1986, Operation Raleigh constructed a Visitor Center and marked the boundaries of the reserve. The area around the Center was fenced, signs were erected, the center was furnished and educational displays constructed.

A Goshen College intern, Emily Fisk, worked at Crooked Tree for six weeks in the Summer of 1986, assessing the tourist potential of the village. Besides this work she screened the Visitor Center and started building a latrine. From 1986 through 1988 Peace Corps Volunteer Jennifer Ellsworth Sanchez was stationed at the Center. Steve Tillett was appointed as the first warden in 1987.

The Visitor Center was built on land leased from John Jex for ten years at the village end of the causeway. When that lease ran out in 1996, the Visitor Center was moved to the other end of the causeway and the displays were improved. But after a continued period of high water from Hurricane Mitch, the Visitor Center was moved back to the village end of the causeway in 1999.

A management plan was developed in 2005 for CTWS by Zoe Goodwin and Paul Walker of Wildtracks.

Ramsar International Wetland

The Ramsar Convention is an intergovernmental treaty that provides the framework for international cooperation in the wise utilization of wetland habitats. The Convention entered into force in 1975 and in 2009 had 158 contracting parties that have so far designated 1,947 areas, totaling nearly 190 million hectares, as "Wetlands of International Importance". On February 26, 1998, the Minister of Foreign Affairs signed the Instrument of Accession to the Convention on Wetlands of International Importance especially as Waterfowl Habitat, known as the Ramsar Convention. CTWS became the first Ramsar site in Belize on August 22, 1998.

Boardwalk

On September 15, 1997, the Belize Audubon Society and the British High Commission inaugurated a new Interpretive Boardwalk for the CTWS. Situated north of the village of Crooked Tree, the boardwalk, the first of its kind in the entire country of Belize, extends for some three quarters of a mile through the lush lowland subtropical forest of this area, alongside the beautiful Crooked Tree Lagoon.

The boardwalk is accessible by small boats during the rainy season, and by vehicular traffic during the dry. This facility is a valuable tool for the Belize Audubon Society's Environmental Education Programme, and for the CTWS wardens to patrol and monitor this area, thereby ensuring the integrity of its flora and fauna. The boardwalk was built through the kind volunteering efforts of Raleigh International, CTWS park personnel, and locals from Crooked Tree Village. Funding was provided by the British High Commission.

CTWS Boardwalk

In 2008 another boardwalk leading from the Visitor Center to Bird's Eye Lodge was constructed with the assistance of the men in the community, funded by Birdlife International.

BAS Archives

Trekforce volunteers on the Observation Tower they built in 1998.

Crooked Tree Lagoon when it is dry.

Observation Tower

On Friday, March 13, 1998, the Belize Audubon Society and a volunteer group known as Trekforce Expeditions of the United Kingdom, officially opened a 25-foot Bird Observation Tower just a few miles south of the populated heart of the Crooked Tree Village. It offers anyone fond of the great outdoors, a spectacular view of the Crooked Tree wetland complex, home to myriads of migratory and resident water birds, and other fascinating wildlife. The Observation Tower also offers the observer a pleasant view of the enchanted Chau Hiix Mayan Site, and makes it an ideal area for education and recreation.

The Role of Wetlands

Sometimes Crooked Tree Lagoon is very dry and sometimes it floods. That is the role of wetlands. When Hurricane Keith struck Belize in 2000, Crooked Tree Lagoon flooded. The causeway was under six feet of water, so access to the village was via boat. That is the role of wetlands, to act like a sponge and absorb all that water. Each time the floods have come, BAS staff helped with hurricane relief.

After Hurricane Dean in 2007, relief was organized by the Belize Red Cross relief with labour provided by the Association of Protected Areas Management Organizations (APAMO) to provide for the needs of all affected communities buffering protected areas. BAS staff participated in this effort.

(*above*) BAS staff helping with boat transportation to and from Crooked Tree village during the floods of October 2008.

(*right*) Peace Corps Volunteer (centre) and BAS Protected Areas Manager Felicia Leal helping with relief after Hurricane Dean.

Crooked Tree Wildlife Sanctuary as Nature School

41st AGM at Crooked Tree Wildlife Sanctuary

Judy Lumb

Osmany Salas

Cockscomb Basin Wildlife Sanctuary

The Cockscomb Basin Wildlife Sanctuary (CBWS) was originally designated as a small no-hunting area within the Cockscomb Forest Reserve for the protection of jaguars. Nick-named the "Jaguar Preserve", the CBWS was declared and then expanded to the current 127,107 acres. Since the beginning BAS has been involved in the establishment and management of the CBWS. Today it is a favorite destination for tourists and Belizeans who enjoy hiking on the miles of trails, floating the South Stann Creek, swiming at the Waterfall, and climbing to Ben's Bluff (*view above*).

Establishment of the CBWS

In the early 1980's concern for the jaguars of Belize was raised from two different places. James Hyde, Permanent Secretary for the Ministry of Natural Resources, had been approached by a citrus farmer who had encountered jaguars in his orchard. Mr. Hyde asked BAS for a study of jaguars. At the same time, Archie Carr III, Assistant Director of the International Division of the New York Zoological Society, ran across references to jaguars in Belize in hunting magazines. He was in contact with Dora Weyer and asked if BAS would like a study of the jaguars of Belize.

Dr. Alan Rabinowitz was assigned to this task and supported by the New York Zoological Society (Wildlife Conservation Society). He first did a country-wide study of jaguars and chose the Cockscomb Basin for his two-year study. At the end of his study in 1984, Dr. Rabinowitz, with the help of Dr. Carr and members of BAS, persuaded the Government of Belize to establish the Cockscomb Basin Forest Reserve and declare it a "no hunting" area (S.I. Nos. 93 and 94 gazetted 24 November 1984). The Forest Reserve was established under the Forest Act of 1976/1981 and the "no hunting" area was under the Wildlife Protection Act.

The World Wildlife Fund wanted to support this project, but they were concerned that the Forest Reserve was not enough protection. When BAS learned that 1,456 hectares had been excluded from the timber concession in the Forest Reserve, they were supported by WWF, CATIE and the New York Zoological Society in their request that this area be designated as a wildlife sanctuary.

BAS Archives

The Cockscomb Basin Wildlife Sanctuary when it first opened in 1986. BAS Executive Director Mickey Craig is second from the left and BAS President James Waight is on the right

BAS Archives

Cockscomb Basin Wildlife Sanctuary Management Workshop sponsored by the World Wildlife Fund, 1987

BAS Archives

The wardens for the Cockscomb Basin Wildlife Sanctuary with visitors (from left) WWF Programme Officer for Central America Steve Cornelius, Pio Saqui, Ignacio Pop, Emiliano Pop, Alfonso Ical, Park Director Ernesto Saqui and an unidentified visitor

BAS Archives

H.R.H. Prince Philip planting mahogany tree in 1988 with Therese Bowman (Rath) and Ernesto Saqui watching

The first National Conservation Conference was held February 26 to 28, 1986, at the Pelican Beach Hotel. Representatives of several international organizations took part in the conference, along with officials of government and representatives of the private sector. After declaring the conference open, Hon. Dean Lindo signed the Statutory Instrument declaring the Cockscomb Basin Wildlife Sanctuary on February 26, 1986 (No. 32 of 1986). This gave that area complete protection and paved the way for funding from World Wildlife Fund-U.S. National Conservation Conference

Management of the Sanctuary

A five-year management plan for the Cockscomb Basin was requested by the Minister. To accomplish this the Cockscomb Basin Management Planning Workshop was held at the Cockscomb Basin Wildlife Sanctuary from March 24 to April 5, 1987. The workshop was sponsored by World Wildlife Fund-U.S. and facilitated by Alan Moore, Wildlands Planner.

Ignacio Pop and his son, Pedro, were hired as the first wardens. Peace Corps Volunteer Dan Taylor, a specialist in wildlife management, was assigned to the CBWS 1986 to 1987. Ernesto Saqui was hired as Park Director in May of 1987 and the staff has continued to grow.

On February 6, 1988, the 19th anniversary of BAS, H.R.H .Prince Philip, International President of the World Wildlife Fund, visited the CBWS. He presented an award to Ignacio Pop and planted a mahogany tree. More than twenty years later that tree has grown over forty feet tall.

Mahogany tree planted by H.R.H. Prince Phillip after 20 years of growth

Expansion of the Cockscomb Basin Wilidlife Sanctuary

Therese Rath

In 1990 the Cockscomb Basin Wildlife Sanctuary was expanded from 3,600 to 102,000 acres. A ceremony was planned for the reserve, but that plan started going awry the day before when continuous rains closed the entrance road to all but four-wheel drive vehicles. Alternate plans were quickly made to hold the signing ceremony at the primary school in Maya Center Village.

The impressive audience included the Minister of Natural Resources, the Hon. Florencio Marin, Senator Conrad Lewis from Hopkins Village, Representative Stanley Usher from Toldeo, Chairmen of all the neighbouring Village Councils, representatives from the two citrus factories, large and small nearby farmers, BTIA representatives and the entire population of Maya Center Village.

But bad weather in Belmopan cancelled the helicopter that was to bring the President of the World Wildlife Fund-U.S., Kathryn Fuller; BAS President, Janet Gibson; BAS Acting Executive Director, Dolores Godfrey and Permanent Secretary in the Ministry of Natural Resources, David Gibson, who was carrying the papers to be signed.

Legal documents were not signed that morning in Maya Center, but anyone who was there will tell you that there was no better way to mark this historic occasion than by having all sectors of the community sitting in a crowded one-room schoolhouse in support of such a momentous event. By 4 pm Minister Marin had made the trip by road back to Belmopan

BAS Archives

Signing of the Statuatory Instrument that expanded the Cockscomb Basin Wildlife Sanctuary (from left) Hon. Florencio Marin, Minister of Natural Resources; Hon. Glenn Godfrey, Minister of Tourism and the Environment; BAS President Janet Gibson; BAS Executive Director Dolores Godfrey; Sharon Matola, Director of the Belize Zoo; Joy Grant, Executive Director of Programme for Belize; Dr. Victor Gonzalez, Chief Environmental Officer

and a small group gathered in his office to witness the signing of Statutory Instrument No. 127 of 1990, Wildlife Sanctuary Declaration (Cockscomb Basin)(Amendment) Order, gazetted on November 17, 1990.

Further Expansion of the CBWS

In 1997 the CBWS was further expanded to 127,107 acres, including the section to the south that connects the CBWS with the Bladen Nature Reserve. This makes a continuous corridor of protected areas totaling 250,000 acres.

Maya Center Women's Craft Group

In 1987 the women of Maya Center organized a craft group. A craft shop was built at the intersection of the road to the CBWS and the Southern Highway where the gate to the CBWS is located. The Maya Center Women's Group is partially responsible for collecting entrance fees and the registration of visitors (*Chapter Seven*).

CBWS as a Nature School. (*back*) Education Manager Olivia Carballo-Avilez, CBWS Park Director Nicacio Coc is the Jaguar mascot and Bonifacio Cal is on the keyboard.

BAS Archives

Bonnie Gestring speaking at the opening of the first CBWS VIsitor Center in 1993.

Judy Lumb

Rosita Pop, Chairwoman of the Maya Center Women's Group, and BAS President Earl Green open the second CBWS Visitor Center in 2009.

CBWS Visitor Centre

On March 29, 1993, the new Visitor Centre in the CBWS was opened. The center was built by villagers of Maya Centre and the excellent displays were the work of Peace Corps Volunteer Bonnie Gestring, who followed the plan designed by artist Deidre Hyde of WWF Costa Rica. BAS Executive Director Virginia Vasquez presented awards to the winners of the poster contest held at the Maya Center School. Music by a Maya Centre marimba enlivened the ceremony.

That Visitor Centre lasted until 2006 when it collapsed. A new Visitor Centre was built through the generous funding from the Jaguar Conservation Fund. A new display was created and officially opened at the 2009 AGM, which was held at CBWS.

New Tourism Facilities

Through a grant from NARMAP, the CBWS developed new, environmentally-friendly tourism facilitaties with composting toilets, and solar-generated electricity and water systems. The new dormitory accommodates 24 persons. Another new building has a central kitchen and dining room. A private cabin was built by Raleigh International volunteers with support from the British High Commission. The new facilities were inaugurated on September 23, 1996. The BAS website has updated information on all facilities and activities.

Judy Lumb

Judy Lumb

Opening of new facilities at CBWS in September, 1996:

(*left*) Raleigh volunteers on the step of the cabin that they built.

(*above*) Bernarda Pop, Celestina Pop holding baby Estevan Pop, and Claudia Saqui in front of the new kitchen at CBWS

(*below*) BAS Secretary Jean Shaw and British High Commisioner Gordon Baker cut the ribbon opening the new dormitory at CBWS.

Judy Lumb

Kaway Tree Swamps on the Wari Trail

Judy Lumb

Judy Lumb

Swimming and Tubing on South Stann Creek

Judy Lumb

Judy Lumb

The view from Tiger Fern Campground

New Trails for CBWS

In 1995 a new backpacker campground called Tiger Fern Campground was developed on the pine ridge overlooking the entire Cockscomb Basin. Subsequently, two waterfalls were found nearby. Also in 1995 Raleigh International made a trail to the Outlier with a backpacker campsite an hour's hike from the peak.

Maya Mopan Entrance for CBWS

In 1997 development of a second entrance was begun for the CBWS. Trekforce volunteers carried out an expedition to the West Basin, starting from the village of Maya Mopan, to begin the process of cutting trails. The next year another group of Trekforce volunteers built latrines at this entrance, but that project was halted a couple of days before completion due to Hurricane Mitch.

Advisory Committee for CBWS

As a result of a village meeting held at the end of October, 1997, in Maya Center in the Stann Creek District, an 11-member Cockscomb-Maya Center Advisory Committee was formed. The meeting brought together about 50 villagers, as well as representatives from BAS, and the Forest and Police Departments. The meeting was called to address concerns presented by the villagers of Maya Center in regard to their relations with BAS and the Cockscomb Basin Wildlife Sanctuary.

The advisory committee was chosen as follows: eight leaders from Maya Center, two BAS representatives and a representative from the Forest Department. The eight Maya center members represent the Village Council, the Women's Group, the two religious denominations, the local school, and the tour guides.

EU-sponsored Co-management Project

The Advisory Committee set the stage for the European Union-sponsored Co-management project from 2000 through 2005 that increased BAS focus on the CBWS stakeholder communities (*for details, see Chapter Seven*).

Victoria Peak Natural Monument

Judy Lumb

Victoria Peak Natural Monument includes the Victoria Peak and the others that make up the Cockscomb (*aerial view above*). These 4,487 acres were declared a Natural Monument in 1998 and have been managed along with CBWS ever since. In 1989 Victoria Peak was still visible from the road in front of the CBWS Park Headquarters (*left*), just a few years after the area had been logged, but it is no longer visible from the Headquarters.

Summary of BAS Management of Protected Areas

Even before it was formed, some of the BAS founders were members of the National Parks Commission, which was commissioned to identify areas in Belize that should be protected. Since its formation BAS has been concerned with the declaration and management of protected areas. The first BAS project in 1969 was to Half Moon Caye to poison rats to protect the Red-footed Booby rookery, which had been declared a Crown Reserve in 1928. As protected areas were declared by the Governmentof Belize (GoB), it was expected that BAS would be involved in their management. Formal agreements were signed between the Society and the GoB to manage six protected areas initially and, eventually, nine protected areas. A ten-year agreement was signed in 2004.

Protected Areas Managed by BAS

In 1982 **Half Moon Caye Natural Monument** was the first protected area declared in the newly independent Belize and then was declared part of the World Heritage Site in 1996.

But the 55-acre **Guanacaste Park** had already been established in 1973 and was declared a Guanacaste National Park in 1990.

Tapir Mountain Nature Reserve (formerly Society Hall Nature Reserve) was established in 1975.

Actun Tunichil Muknal Natural Monument was declared in 2004 to protect the cave in the Tapir Mountain Nature Reserve and allow visitation.

Crooked Tree Wildlife Sanctuary was declared in 1984 and became an internationally protected wetlands under the Ramsar Convention in 1998.

Cockscomb Basin Wildlife Sanctuary and **St. Herman's Blue Hole National Park** were declared in 1986.

Blue Hole and Victoria Peak Natural Monuments were added to BAS-managed protected areas in 1999. They protect the famous Blue Hole in Lighthouse Reef, a collapsed cave ceiling, and the Cockscomb Mountain Range.

Belize Audubon Society

Celebrating 40 Years

Chapter Five

Research for Biodiversity and Conservation

BAS Research for Biodiversity and Conservation

The overall goal of the BAS Research Programme is to provide scientific information to protected areas managers to enable them to develop targeted management actions that ensure the continued viability of the ecosystems and species that the protected areas aim to conserve. Research may be initiated by BAS, by scientists living in Belize, or by scientists or students from abroad. No matter how the research is initiated, research permits from the Forest Department are required.

The Belize Audubon Society has been involved in research throughout its 40-year history. Research was carried out by volunteers or visiting foreign scientists until 2000 when Christopher Hecker was hired as Natural Resources Data Manager through a joint program with Wildlife Conservation Society. Roberto Pott succeeded him as Research Coordinator in 2003 and then Ivis Chan took the position in 2005.

Christmas Bird Counts

Ford Young

From the first year BAS has conducted Christmas Bird Counts and reported the results to the National Audubon Society. A Christmas Bird Count (CBC) is established in an area that is a 15-mile circle around an established point. The CBC occurs over a 24-hour period within two weeks including the Christmas Holidays. A compiler organizes each count. The compiler collects the tallies from each party, compiles the data and submits it to the National Audubon Society <audubon.org/bird/cbc>. The report includes weather condition, number of parties, and names of participants, as well as the total party hours. Initially both the number of species and the total number of birds were reported. After 1998 only the total number of species were reported.

Ford Young was the first BAS compiler and remained compiler through 1986, when the job of compiler was then taken over by Lydia Waight and Meg Craig, followed by Dr. Douglas James, and Micky Craig, with Philip Balderamos and Dirk Francisco as secondary compilers.

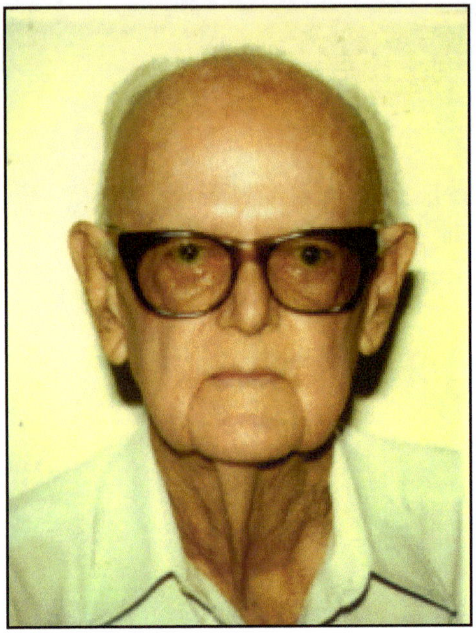

W. (Walker) Ford Young

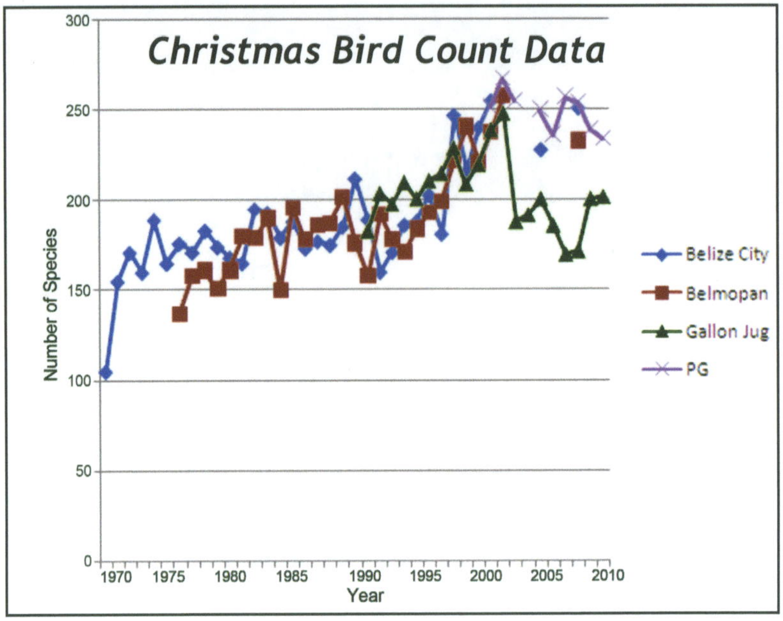

Christmas Bird Count Data 1969–2009

The Belize City count area is centred about three miles north 34 degrees east of the village of Burrell Boom and includes parts of the old Northern Highway, sea coast, Bermudian Landing Road, Mussel Creek, Belize River, International Airport, Ladyville, Sand Hill and Hattieville. The highest number of species counted in the Belize City area was 257 in 2001.

In 1975 the second CBC, the Belmopan count area, was established with the centre 2.25 miles east of mile post 49 on the Hummingbird Highway. It includes Belmopan, Guanacaste Park, Caves Branch and Sibun cacao plantation, the Western Highway from Beaver Dam Creek to Camalote, parts of the More Tomorrow and Never Delay Roads, and the Hummingbird Highway from the Western Highway to Silver Creek. The initial count was 137 species.

A third Christmas Bird count, with its centre at the village of Gallon Jug, was established in 1990 by Bruce Miller, compiler. It includes Laguna Seca, the south end of the Booth's River escarpment, Sylvester's camp and the north end of Chan Chich Creek. Except for a cleared area of about 2,000 acres around Gallon Jug, the balance consists of Tropical Moist Forest with access on foot along old logging trails opened by Belize Estate and Produce Co. Ltd., many years ago. This count area has the advantage of supporting many forest birds not normally found in the other two count areas. The number of species recorded in this count area has varied, from 170 to 248.

In 2000 a fourth Christmas Bird Count Circle was established in Punta Gorda led by H. Lee Jones. As can be seen in graph above, the numbers of species has been high from 236 to 267.

In 2008 BAS started a new count in the Cockscomb Basin. Dirk Francisco is the compiler for the Cockscomb Basin CBC. The initial count was 139.

Important Bird Areas and Christmas Bird Count Areas

Carolyn and Bruce Miller

The Belize Important Bird Areas Project was funded by BirdLife International through BAS as the local partner. The important bird areas (IBAs) are critically important bird and biodiversity sites that, in combination, encompass all major ecosystems.

Methods used were those standardized across the earth to identify IBAs on the basis of bird numbers and species' complements that meet one of the following triggering criteria:

- Species of global concern
- Species with a restricted range of distribution
- Species restricted to a certain broad habitat type
- Large congregations of birds

The IBA Program also aims to guide the implementation of national conservation strategies through the promotion and development of national protected-area programs. When possible, IBAs are to be linked to existing protected areas. The six IBAs identified are shown on the map (*right*) in relation to the five Christmas Bird Count areas and listed on the next page.

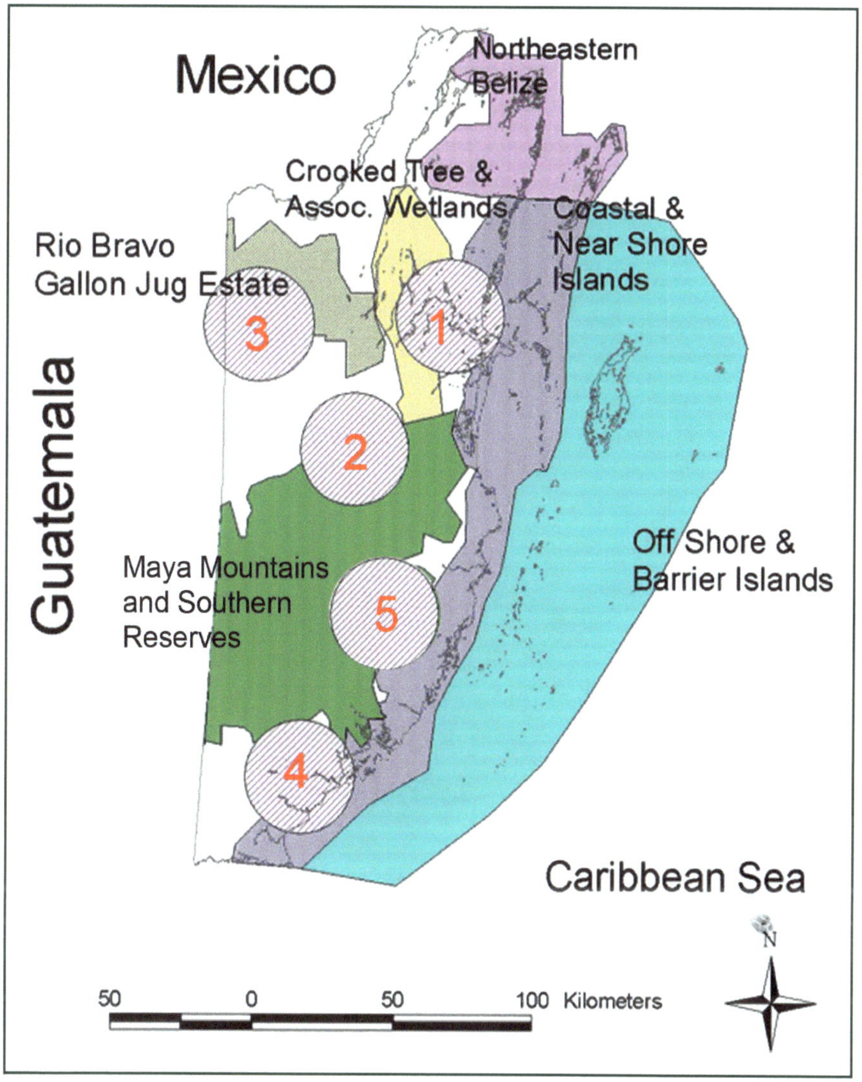

Christmas Bird Count Area and Important Bird Areas

Christmas Bird Count Areas

#	Count Area	Year Established	Average # species
1	Belize City	1969	189
2	Belmopan	1975	189
3	Gallon Jug	1990	204
4	Punta Gorda	2000	249
5	Cockscomb	2009	139

Northern Jacanas (Jacana spinosa) in Crooked Tree Lagoon

Identified Important Bird Areas

- **Rio Bravo and Gallon Jug Estate** includes Aguas Tiubias National Park
- **Crooked Tree and Associated Wetlands**
- **Northeastern Belize** includes Shipstern Nature Reserve, Fireburn, Freshwater Creek Forest Reserve, Bacalar Chico and Honey Camp National Parks, Cerros Maya Archaeology Reserve, Corozal Bay Wildlife Sanctuary, and two Bird Sanctuaries: Doubloon Bank and Little Guana Caye
- **Maya Mountains and Southern Reserves** includes most of the terrestrial protected areas in Belize.
- **Coastal and Near Shore Islands** includes three Marine Reserves: Hol Chan, Caye Caulker and Port Honduras; and seven Forest Reserves: Grants Works, Mango Creek, Swasey-Bladen, Machaca, Caye Caulker, Deep River and Manatee; four Wildlife Sanctuaries: Aguacaliente, Gales Point, Swallow Caye, and Corozal Bay; two National Parks: Sarstoon-Temash and Payne's Creek; four Bird Sanctuaries: Bird Caye, Monkey Caye, Los Salones and an un-named caye; Altun Ha Archaeological Reserve and Burdon Canal Nature Reserve.
- **Offshore and Barrier Islands** includes five Marine Reserves: South Water Caye, Gladden Spit, Silk Cayes, Sapodilla Cayes and Glovers Reef; Half Moon Caye Natural Monument, Laughing Bird Caye National Park and Man of War Caye Bird Sanctuary.

OTHER BIRD COUNTS AND FESTIVALS

In addition to the Christmas Bird Counts, other special bird festivals and counts have been held throughout BAS history, usually in October or April when large numbers of migrants are present. Although some of these events were for fund-raising purposes, they always provide useful information on the distribution of birds in Belize.

October Bird Festivals

The first BirdFest International was held October 2 to 8, 1998. It was designed to be a major fund-raising event for BAS, taking advantage of off-season rates in accommodations, combined with the large numbers of migrating birds that are in Belize in October, along with the resident species. Four different itineraries were offered for a maximum of 150 participants. Participation was very low, but all were quite pleased with the experience. In all, 175 species of birds were seen.

World Bird Counts sponsored by BirdLife International and other international bird organizations ask birders to count species seen in the month of October. BAS has promoted these counts and compiled the results. In 1999 the count was higher, 378 species.

In the wake of Hurricane Keith, Belize BirdFest 2000 welcomed visiting birders from sister birding clubs and Audubon Societies in the U.S. For eight days, eight birders were escorted by Glen Crawford of Crooked Tree all over the country: Northern, Western, Hummingbird, and Southern Highways; Crooked Tree and Cockscomb Basin Wildlife Sanctuaries; Dangriga, South Water Caye, and Belize City. They identified 234 species, including six species that were listed as "rare" in the BAS-published bird checklist by Miller and Miller.

Magnificent Frigatebird (*Fregata magnificens*)

Judy Lumb

Boat-billed Heron (*Cochlearius cochlearius*)

Osprey (*Pandion haliaetus*)

American Oystercatcher (*Haemotopus pallitus*)

On 1st October 2006 BAS Bird Fest was held at St. Johns College, the Fort Area, and Birds Isle in Belize City where 50 species were seen.

As part of BirdLife's World Bird Festival celebrated from October 6-7, 2007, BAS assisted the West Landivar Citizens Action Committee in conducting a rapid ecological assessment of the birds in the newly formed protected area of Dolphin Park. A total of 37 species were seen.

BAS celebrated the World Bird Festival throughout October of 2009 with a series of urban bird walks in San Ignacio (49 species), Dangriga (66 species), and Corozal (59 species). As avid birders know, the month of October is exciting as many birds begin their winter migration.

April—June Bird Festivals

As part of the BAS 30th anniversary celebration in 1999, the first Neotropical Birdathon was held throughout Belize. Four teams participated. In Crooked Tree Wildlife Sanctuary the teams saw 130 species; in Cockscomb Basin Wildlife Sanctuary two teams observed 166 species; and the fourth team saw 121 bird species from Belize City to Hill Bank. The grand total was 242 species of birds.

The Neotropical Birdathon grew in 2000 to 12 teams that observed more than 350 species.

The Neotropical Birdathon was a fund-raising event in 2001 that took advantage of the large number of bird migrants passing through Belize in April. In partnership with the National Audubon Society, supporters abroad made pledges to contribute a certain amount per bird species seen in one defined 24-hour period, 6 pm April 21st to 6 pm April 22nd. Birders all over Belize went out with their binoculars and sent in their results to the BAS Office. A total of 265 species were seen.

Red-footed Booby Studies

The BAS founders were concerned about the Red-footed Booby colony on Half Moon Caye. In the 1970s, volunteer Goshen College students conducted studies of the Red-footed Boobies on Half Moon Caye for several years. No study had been done on Half Moon Caye since the one by Jared Verner in 1958 in which he estimated 3,500 boobies and counted 1,389 nests. Considerable damage had been done by Hurricane Hattie on Half Moon Caye in 1961 and BAS wanted to know if the colony had been restored to its previous levels. In November of 1974, Goshen students estimated 3,700 birds, one-sixth of which were immature. The next March 1,285 nests were counted and in 1977 there were 1,431 nests.

During the Christmas holidays of 1991 William E. Cross of Environmental Research Associates in Ontario, Canada, did another study of the nesting colony on Half Moon Caye. He counted 1325 Red-footed Booby nests and 71 Magnificent Frigatebird nests. He also observed iguanas (*Iguana iguana rhinolopha*) and wish-willies (*Ctenosaura similis*), which were present in about every fourth tree.

Judy Lumb

(*above, left*) Red-footed booby (*Sula sula*) chick,

(*above*) Red-footed booby adult sitting on nest

(*below*) Magnificent Frigatebirds (*Fregata mangificens*) in all their magnificence

Judy Lumb

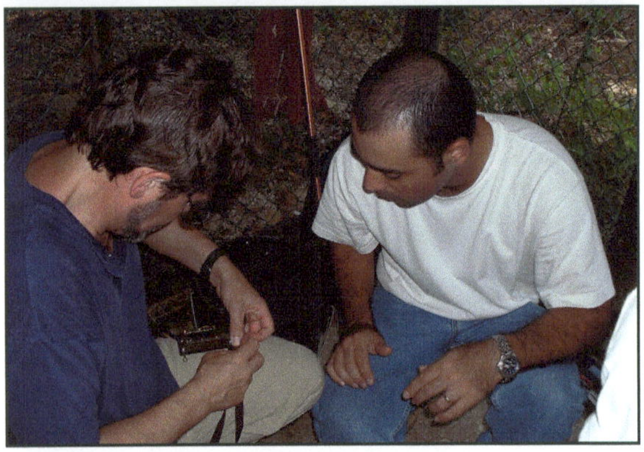

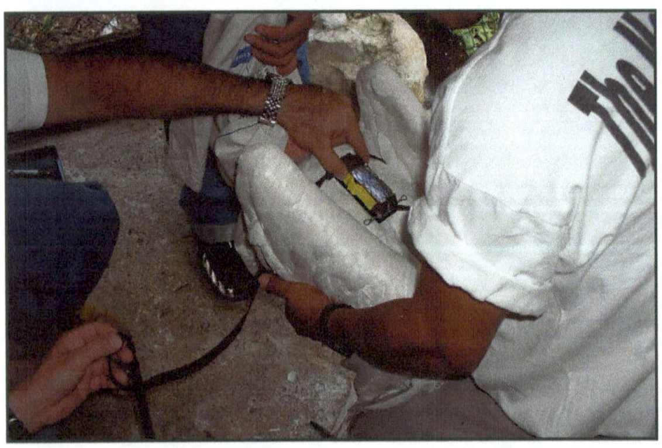

(*top*) Marcos Cucul with a Jabiru nest at Hill Bank (*centre*) Dr. Ken Meyer and Omar Figueroa (*bottom*) GSP/satellite transmitter is attached to a young Jabiru. (*photos courtesy of Omar Figueroa*)

Jabiru Studies

W. Ford Young

BAS began research on Jabirus in the very first year when local pilots reported seeing nests with large black and white birds in northern Belize. BAS Vice President W. Ford Young was a pilot, so he flew over the area and confirmed the two Jabiru nests. He noted that the two nests were miles apart in large dead trees built of sticks and twigs that formed a platform roughly 8 to 10 feet across with each nest containing two fledglings.

BAS then undertook the task of protecting Jabirus (*Chapter Two*). Since that time Audubon members and employees of the Society in the Crooked Tree area have encountered increasing numbers in the Crooked Tree Lagoon (also known as Northern Lagoon). This culminated in a count in late May, 1993, of 50 birds. Based on increased numbers of bird flocks preparing to migrate, reports of feeding birds from the general public, and the number of nests located, it is suggested that the number of birds in Belize has increased from around 20-30 in the early 1970s to approximately 60-70 in 1993.

Alejandro Paredes

Alejandro Paredes of the University of Florida carried out surveys and counted 59 Jabirus, half of which were observed in front of the Crooked Tree Wildlife Sanctuary Visitor Centre.

Omar Figueroa

Former Director of Birds Without Borders Omar Figueroa studied Jabirus in Belize as part of his work for a Master of Science degree at the University of Florida. In 2003 ten Jabiru nests were identified, eight of which were in protected areas, including Crooked Tree Wildlife Sanctuary. In 2003 three young Jabirus were tagged with GPS/satellite transmitters. In the first few months after they fledged, the bird stayed within one kilometer of their natal nest.

Bird Observations in the BAS Newsletter

The Editor of the *BAS Newsletter* is a lurker on the Belize Birders' email listserv and finds many observations that warrant an article in the *Newsletter.* Here a few of the many reports:

Rudi Burgos—White Great-Tailed Grackle

Following several reports of a "white" blackbird reaching BAS, Dirk Francisco and I visited the High School at Ladyville. From conversations with the security guard staff, some teachers, and a few students, this bird has been around for some time. The grackle surprised us by suddenly appearing shortly after we arrived almost causing me a heart failure!

Albino Great-tailed Grackle (*Quiscalus mexicanus*)

Lee Jones—Roseate Terns in Punta Gorda

On 2 December 2007 I saw approximately 60 Roseate Terns and two Common Terns in with Royal, Sandwich, and Laughing Gulls on the customs wharf in Punta Gorda. According to the literature, the Roseate Tern is not supposed to be in the western Caribbean in winter, but since this is essentially a pelagic (i.e., offshore) species, we really don't know much about its seasonal distribution compared with species we can typically see from land.

Philip Balderamos—Canada Goose in Belize

On 9 November 2007 I was surprised to see a Canada Goose in the Crooked Tree Lagoon.

Mark Ware

Canada Goose (*Branta canadensis*)

Wil Maheia—Sooty Tern Nesting Site

I was just out at the Snake Cayes and saw the Sooty Terns coming around the first of March. I believe this is the only nesting site for them in Belize. If so, I will need all your help as the Caye they are nesting on was leased out after the declaration of the Port Honduras Marine Reserve. I hope that we can get the Minister to cancel the lease.

Jan Meerman—Black and White Hawk-Eagle

In mid-February, 2007, I observed an unfamiliar raptor a couple of hundred yards away from my home. I grabbed binoculars and camera and soon it became obvious that it was a Black and White Hawk-Eagle! Not the most common of sightings!

Wayne Hall—Keel-billed Motmots

We saw Keel-billed Motmots at two locations in Mayflower-Bocawina National Park, near the Tau Witz Trail and along the Bocawina Falls Trail where it first approaches Silk Grass Creek.

BAS Archives

Harpy Eagle (*Harpy harpyja*)

BAS Archives

Collared Aracari(*Pteroglossus torquatus*)

BAS Biodiversity Monitoring

Roberto Pott and Michael Link

The objective of this monitoring programme is to provide baseline data on which the ecological integrity of the various protected areas can be gauged using birds as an indicator. This bird monitoring follows the recommendations for monitoring regimes designed for landscapes and species of the Selva Maya Region, which includes Belize, the Peten Area of Guatemala, and the Yucatan Peninsula of Mexico.

BAS piloted its Biodiversity Monitoring Program at two of its protected areas, Crooked Tree Wildlife Sanctuary (CTWS) and Blue Hole Natural Monument (BHNM) in April and May of 2003. With the support of Wildlife Conservation Society, training in new monitoring methodologies was conducted for birds and mammals. Sites were established in both CTWS and BHNP. Staff members Rennie Jones and Steve Tillett (CTWS), and Israel Manzanero and Dennis Williams (BHNP) recorded their observations, which were compiled by BAS Research Coordinator Roberto Pott and processed by BAS Data Manager Michael Link.

The monitoring programmes was expanded to four protected areas, Blue Hole National Park (BHNP), Cockscomb Basin Wildlife Sanctuary (CBWS), Crooked Tree Wildlife Sanctuary (CTWS) and Tapir Mountain Nature Reserve (TMNR). Two sites, one pristine and one disturbed, were selected for each protected area. There are two monitoring periods within a given year: 1) January/February and 2) May/June. Counts were conducted along a two-kilometer transect at 100 meter intervals. Each point was limited to a radius of 25 meters and included auditory and visual observations.

Migrant species made up a significant portion of species recorded for CTWS transects A and C with migrant percentage of species recorded being 21% and 28%, respectively, and BHNP Transect B had a percentage of 19%. A new species recorded for the BHNP was the Gray-throated Chat (*Granatellus sallaei*) and the Keel-billed Motmot (*Electron carinatum*) was observed within CBWS.

There are several species that occur within Belize that are of national and regional concern due to habitat destruction and hunting throughout the region such as the Crested Guam (*Penelope purpurascens*) and the Great Curassow (*Crax rubra*). These species have been documented in several of the parks in the past. Although not seen during the survey, they are repeatedly observed in some parks, especially CBWS. During monitoring efforts in June 2004, we observed both Crested Guam and Great Curassow at BHNP.

BAS Archives

Ivis Chan

Under the Darwin project, bird surveys were conducted at CBWS and CTWS during December 2006 and January 2007. These surveys were conducted by Dr. Lee Jones along with Israel Manzanero Sr. from St. Herman's Blue Hole. University of Belize students Emmerson Garcia and Marvin Vasquez, and Normando Mora, the Darwin project field technician, also formed part of the team. Field work completed for the Darwin project at CBWS included GPS reading, canopy height, soil pH, and tree density.

Members of the UK Center for Ecology and Hydrology, conducted introductory training in the use of the CANOCO package for multivariate analysis.

James Beverdge

Blue-crowned Motmot (*Momotus momta*)
Vermillion Flycatcher (*Pyrocephalus rubinus*)

Research in Crooked Tree

Tom Meekin

Peace Corps Volunteer Tom Meekin, the husband of Catherine Meekin who worked in the BAS Office, made a fishery survey of the Crooked Tree Lagoon in 1985.

University of California at Davis

Researchers from the University of California at Davis studied wetland communities in northern Belize from 1993 to 1995. They analysed plant, water and soil samples. They found that the dominant plant species are cattails (*Typha domingensis*), saw-grass (*Cladium jamaicense*), and rushes (*Eleocharis cellosa* and *E. interstincta*).

Water Quality Monitoring

Volunteer Woodard Miley worked at Crooked Tree May 26 to June 7, 1994. He established a baseline water quality monitoring programme, a recording procedure for recreational and commercial fish catch, and made additions to the draft management plan. His work was conducted under the Farmer-to-Farmer Programme of the Florida Association of Voluntary Agencies for Caribbean Action. Equipment was provided to the Sanctuary and personnel were trained in its use. Water quality parameters are now being recorded annually.

Manatee Studies—ECOSUR

El Colegio de la Frontera Sur (ECOSUR) has studied manatees in Chetumal, Mexico, since 1990. A study about manatee demography, health and genetics was carried out from 2004 to 2007. GPS satellite telemetry technology was used to tag and track manatees. The last of the tagged males, called "Manchas" because he had a big pink patch on his back, moved to the Belize River and into Northern Lagoon in the Crooked Tree Wildlife Sanctuary where he remained for 45 days (*map at right*).

BAS Archives

Donald Tillett (left) and Rennie Jones testing the water in Crooked Tree Lagoon

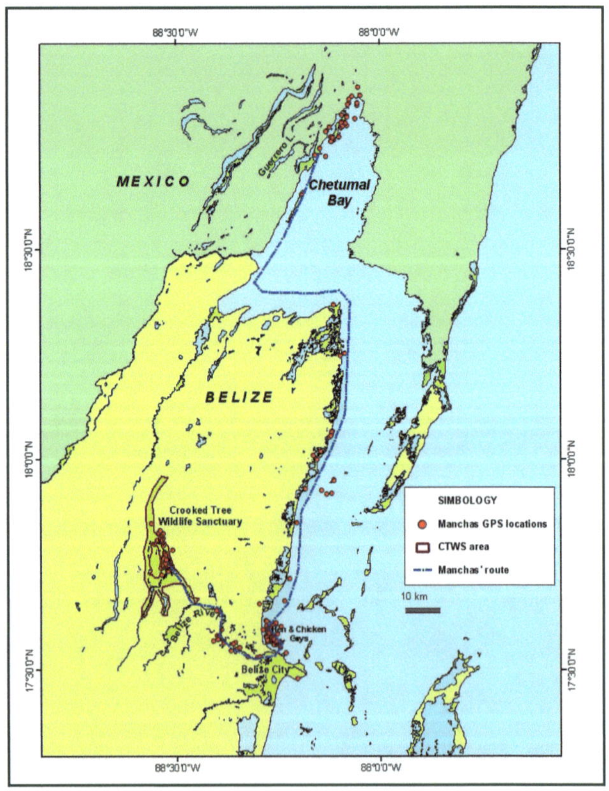

Coutesy ECOSUR

Ed Boles and Pio Saqui, University of Belize

In 2003 students of the Natural Resource Management Programme of the University of Belize (UB) under the supervision of UB Lecturers, Dr. Ed Boles and Pio Saqui, conducted a Rapid Ecological Assessment of biological and socio-economic systems of the Crooked Tree Wildlife Sanctuary. A team of limnology students conducted basic water quality tests in the lagoon.

The surveys revealed the possible association of algae with a thick sediment layer. This may contribute to the successful colonization of the Crooked Tree Lagoon by tilapia, which feed on algal muds.

The assessment also included impact mapping using methods that require minimum scientific research capabilities. A survey of impacts continue to display very high impacts around the communities of Lemonal and Crooked Tree that are linked to resource use in these areas.

Surveys reported an increase in the frequency of large Tarpon (*Tarpon atlanticus*). Tilapia population continues to increase within the lagoon. One species apparently affected by tilapia is the Mos Mos (*Cichlasoma spp*.), and, while an explanation is not obvious, it suggests an opportunity for further investigation. The fisheries continue to supply small scale local markets, mostly to Orange Walk Town, with tilapia fillet fetching as much as six dollars per pound and fishers report that catches have remained consistent over the years.

The methods used during this study are applicable to other communities in Belize. The research team, hopes to develop a manual of low-tech methods that may be adopted by communities for implementation in their respective freshwater systems on which their livelihoods depend.

Crooked Tree Wildlife Sanctuary
Steve Tillett and Donald Gillett, Jr.

Crooked Tree Lagoon

Research in Cockscomb

The Cockscomb Basin has attracted many scientists, beginning with the first zoological expedition by Blake and Agostini in 1935. Stephen M. Russell collected birds in 1958 and 1959. James Kamstra did a complete ecological survey of the Cockscomb Basin in 1986. His Master's thesis has been used ever since by scientists working in the CBWS. Some research projects are listed below.

Research in Cockscomb Basin Wildlife Sanctuary

Researcher	Subject of Research
Timothy McCarthy, M.L. Reed and W. B. Burton	small mammals
Michael Konecny	smaller cats
Daniel Taylor	reptiles
Daniel Taylor	bats
Katherine M. Emmons	environmental education
Jan Meerman and Tineke Boomsma	butterflies, dragonflies and damselflies
Carol Schutte and John Shuey	butterfly and beetles
Philip Elliot	netted and banded birds
Christina Colon	attitudes in Maya Center
Jonathan Lyon	phenology
Jeremy F. Jacobs	mammals
Adrian Barnette and Rebecca Shapley	bats
Patrick Kangas and Ed Boles	fish
Jake and Kelly Marlin	frogs
Laura Marsh and Rebecca Foster	mammal
Bruce and Carolyn Miller	bats
John Shuey	butterflies

BAS Archives

Tracking black howler monkeys in Cockscomb in 2003, Valdemar Andrade (*second from left*), Ernesto Saqui (*right*)

Judy Lumb

Still on display in CBWS, these traps were used by Alan Rabinowitz for his jaguar research in 1983–1984.

Cockscomb Basin Mammal Survey

Scott C. Silver and Linde E. T. Ostro, WCS

This mammal survey was conducted in the Cockscomb Basin Wildlife Sanctuary between May 26 and June 22, 1999. The research team consisted of five biologists, three Field Assistants from Maya Center, and two volunteers from NewYork. Mammals were detected by several different methods:

1) **Camera traps** were placed on either maintained trails or unmaintained trails one-half meter from the ground and set to take pictures at one minute intervals. Nine mammals were identified.

Camera trap results

Nasua nasua	Coati
Agouti paca	Gibnut
Dasypus novemcinctus	Armadillo
Mazama americana	Brockett deer
Conepatus semistriatus	Striped hog-nosed skunk
Felis pardalis	Ocelot
Felis wiedeii	Margay
Felis concolor	Puma
Panthera onca	Jaguar

2) **Live small mammal traps** were placed runs, holes, hollow logs, on the trunks or limbs of trees and on tops of fallen tree trunks. Nine additional mammals were identified.

Small mammals trapped

Oryzomys alfaroi	Rice rat
Oryzomys coucsi	Rice rat
Oryzomys rostratus	Rice rat
Sigmodon hispidus	Hiispid cotton rat
Philander opossum	Grey four-eyed opposum
Marmosa mexicana	Mexican mouse opposum
Ototylomys phyllotis	Big-eared climbing rat
Heteromys desmarestianus	Spiny pocket mouse
Didelphis marsupialis	Common opposum

3) **Tracks** seen on the trails after rains included jaguar, puma, ocelot, brockett deer, collared peccary, gibnut, agouti, armadillo, coati, and tapir.

4) **Night walks** identified six additional species: kinkajou (*Potos flavus*), prehensile tailed porcupine (*Coendou mexicanus*), water opossum (*Chironectes minimus*) and three species of arboreal opossum (*Didelphus viginana, D. marsupialis, Caluromys durbianus*).

Bruce and Carolyn Miller

5) **Mist nets** were used to collect bats. Twelve species were identified. One species, *Thyroptera tricolor* was caught by hand from the rolled leaves of heliconia plants.

Bats captured by mist nets

Noctilio leporinus	Fishing bat
Micronycteris schmidtorum	Schmidt's big-eared bat
Saccopteryx bilineata	Greater white-lined bat
Chrotopterus auritus	Wooly false vampire bat
Glossophaga sp.	Long-tongued bat
Carollia brevicauda	Silky short-tailed bat
Carollia perspicillata	Seba's short-tailed bat
Sturnira lilium	Little yellow-shouldered bat
Artibeus literatus	Grey fruit-eating bat
Artibeus phaeotis/watsoni	Fruit-eating bat
Artibeus toltecus	Toltec fruit-eating bat

The current survey recorded 72 mamml species, bringing the total known for the reserve up to 89. Including the bat species identified acoustically and harp-trapped by the Millers, a total of 22 mammal species were added to the list of species known to occur in the CBWS. New additions included two marsupials (*Didelphis virginiana, Caluromys derbianus*), one insectivore (*Cryptotis purou tropicalis*), one rodent (*Oryzomys couesi*) and 18 chiropterid species.

Courtesy WCS

Ocelot captured by a camera trap

Jaguar Research

Alan Rabinowitz, WCS

Courtesy WCS

Mark Wechsel and Bart Harmesen

The Wildlife Conservation Society (WCS) supported the jaguar study of Alan Rabinowitz that led to the establishment of the Cockscomb Basin Wildlife Sanctuary (CBWS) from 1983 to 1984. Ben Nottingham assisted Alan Rabinowitz and did his own ecological studies. They captured five jaguars by treeing them with dogs or baiting large steel traps with pigs. Radio collars were attached so the jaguars could be tracked. The results indicated that the jaguars were active just after sunset and throughout the night. From scat that was gathered, Rabinowitz learned that over half of the jaguars' meals in Cockscomb were armadillo, as well as some gibnut and anteater. Home ranges for the males were 25-38 square kilometers and half that for the females. In a follow-up study in 1991, Rabinowitz estimated that there were between 600 and 1,000 jaguars in Belize.

Supported by WCS and under the direction of Linde Ostro and Scott Silver, jaguar research continued in CBWS from 2000 through 2009. Baalum Naj, the cabin that housed the original research by Alan Rabinowitz was renovated to house the researchers and their equipment and materials, including scat collections.

The research used camera traps instead of a steel traps which is much less invasive. Much work had to be done to test this new methodology. Because of their spots, individual jaguars could be identified from the photos. Using camera trap technology, 77 individual jaguars have been identified in and around the park since 2000, providing information about density, habitat use, spatial distribution and social structure. On-going scat collection has contributed over 500 samples to study jaguar and puma diet in the area. CBWS jaguar density estimates ranged from 6.3–9.3/100 square kilometers.

Silverio Y. Marin, Bart J. Harmesen
Paul Higginbottom

Methods complementary to the camera trap were investigated for jaguar population assessment. Results indicated a density of 11 jaguars per 100 square kilometers.

Said Gutierrez and Bart Harmsen

Another ancillary method took advantage fo the jaguar's scrape marking behaviour. Through the use of photographic captures, individual jaguars within an area of approximately 125 km^2 were identified. and scrape marks on two major trails were recorded over a period of two months. Through the analysis of the data, an explanation of the purpose of the scrape marking behaviour in jaguars was formulated. It is concluded that jaguars in the CBWS scrape as a sign of immediate temporary occupation of an area.

Rebecca Foster

In 2004 the shooting of a jaguar at a cattle ranch near the CBWS prompted a study documenting cattle kills by jaguars and other large predators. WCS sponsored a Jaguag Predation Workshop for BAS, the staff of the Forest Department, and the cattle ranchers in the area aimed at differentiating between jaguar and puma kills and sharing information and concerns. The Workshop included a site visit to one of the nearby ranches to assess the situation and further the discussion on how it can be addressed jointly by all the relevant stakeholders.

Rebecca Foster and Bart Harmsen

Because of the increase in cattle ranches that have been carved out of the rainforest and the resulting concern about jaguar predation, studies were instituted to compare the activity and predation of pumas and jaguars. Jaguars and pumas had similar nocturnal activity schedules and used similar habitats in CBWS. However, they avoid using the same area at the same time and their diets were quite different. In the rainforest jaguars eat mainly nine-banded armadillos (*Dasypus novemcinctus*), as was shown in the Rabinowitz study in the 1980s. Pumas eat mainly gibnut (paca, *Agouti paca*). Both took larger prey in the forest. for jaguars it was white-lipped peccaries (*Tayassu pecari*) and for pumas, red brocket deer (*Mazama americana*).

Camera traps were also set up near cattle ranches. Pumas were rarely seen outside the forest, but jaguars were often found in areas of human habitation. In unprotected fragmented lands, jaguar scats rarely contained large wild prey species, but showed a diet of small wild prey supplemented with larger domestic species.

Jaguars caught in camera traps in CBWS. (*photos courtesy WCS*)

Judy Lumb

Black Howler Monkey Translocation

Fred Koontz (WCS), Ernesto Saqui, Rob Horwich (CCC)

The reintroduction of black howler monkeys (*Alouatta pigra*) into the Cockscomb Basin Wildlife Sanctuary (CBWS), a joint project of the Belize Audubon Society, Community Conservation Consultants (CCC) and the Wildlife Conservation Society (WCS), was conducted in May, 1992, 1993, and 1994. Fred Koontz of the WCS was the Project Directer, along with Ernesto Saqui, CBWS Park Director and Rob Horwich of CCC. Each year the team consisted of persons from the BAS, CBWS, the WCS's Bronx Zoo and the Community Baboon Sanctuary (CBS) villages, who helped capture the monkeys and graciously donated them to the CBWS. The British Armed Forces provided a helicopter to transport some of the animals.

Black howlers were living in the Cockscomb Basin as recently as the 1960's, but a combination of yellow fever in 1956-7 and habitat destruction by Hurricane Hattie in 1961 probably accounted for the local demise of the black howler.

After the first year, troops were followed in the CBS, prior to the translocation so all members could be identified. Each howler in the troop was darted with a chemical immobilization drug and caught in a net as it dropped to the ground. Each animal was weighed, measured and given health checks by Dr. Wendy Westrom, the team veterinarian. She took blood samples so that genetic records and blood chemistries of the founding population would be available for future research. Each adult animal was fitted with an anklet with colored beads for individual identification and radio transmitters so they could be monitored by tracking them with a directional radio receiver.

(*top*) Black Howler Monkey
(*bottom*) Acclimation cage in CBWS

They were first released into acclimation cages at the release sites in the CBWS to make sure the animals were healthy, to observe the monitoring equipment, and to give them a chance to recover from the capture. In all, 62 animals from 14 troops were reintroduced into the Cockscomb Basin between 1992 and 1994. Hermelindo Saqui, Emiliano Pop, and Pedro Pixabaj tracked the reintroduced monkeys for a year after the last translocation. Their work provided much valuable information on their behavior and ecology.

The translocated black howler monkeys adapted well to the CBWS. A year after the last translocation, only two of the translocated animals were known to have died and 48 could still be located. Monkeys mostly stayed close to their release sites, with the average dispersal being less than 2 kilometers. Eighteen babies had been born in the CBWS and at least 14 of them were still alive. The total population was estimated between 55 and 75 animals.

A final census in 1997 revealed that the population suffered very low mortality and was increasing at an estimated rate of 14.7 percent per year. Ten years later in 2007 a census of black howler monkeys in CBWS by Linde Ostro (Wildlife Conservation Society), Scott Silver (Queens Zoo), Emiliano Pop and Eligorio Sho revealed that 21 different groups were living in the CBWS and the surrounding areas.

Some groups showed remarkable fidelity to home ranges. Harold's Group, the United Group, and the White House Group were found in precisely the same areas, using the same travel paths as when they were followed in 1994–1995. In addition, groups were identified in many new areas during this survey, indicating that monkeys are dispersing down the Access Road and along South Stann Creek and the streams that flow into it, establishing groups in new areas throughout the Basin.

Judy Lumb

Judy Lumb

(*top*) Scott Silver and Fred Koontz carry a transport cage with a monkey inside as Ernesto Saqui watches.

(*bottom*) The translocation team discusses the radio tracking of the translocated monkeys. (*from left*) Ken Glander, Rob Horwich, Fred Koontz, Hermilindo Saqui, Scott Silver, Ernesto Saqui, and Wendy Westrom

Victoria Peak Amphibian and Biodiversity Expedition

Ivis Chan

The foolhardy team included BAS staff Marcelo Pau and Ivis Chan, Wildtracks director Paul Walker and botanist Zoe Goodwin, a group of youngsters from surrounding communities, and Ruth Akers, a graduate student from the University of Edinburgh. The purpose of the trip was to collect data for the Darwin Initiative, to collect botanical specimens from the summit of Victoria Peak and to record sightings of upper elevation amphibian species.

We collected data for the Darwin Initiative in the vicinity of 14K. For our hard work, we were rewarded on different occasions by a rare sighting of the Keel-billed Motmot (*Electron carinatum*), and a pair of charismatic Scarlet Macaws (*Ara macao*) who perched just long enough to be photographed. CBWS warden Marcelo Pau sighted a beautiful juvenile Crested Eagle (*Morphnus guianensis*).

Finally the day came to begin our ascent to Victoria Peak. The first part of the climb that morning was a nightmare of ups and downs seconded only by the latter part of clambering on precipitous ledges, and finally being hoisted up on a rope. Once past the rope, we entered the "Submontane Elfin Shrubland" ecosystem. The ground was covered with sphagnum moss, and the canopy of trees was no more than two to three meters high. At the very summit, however, the vegetation of primarily shrubs and moss was bare to the sun and elements. From this vantage point we could appreciate views of the carpet of forest stretching out below in every direction. We also had a striking view of Outlier Peak and some of the other peaks that make up the Cockscomb range.

Ivis Chan

Crested Eagle (*Morphnus guianensis*)

Ivis Chan

Victoria Peak Expedition Participants

(front center) **Marcelo Pau** *(center row from left)* **Zoe Goodwin, Ruth Akers, Adolfo Pop, Santos Tush, Ivis Chan, Alejandro Bolon;** *(back row from left)* **Renny Pau, Sebastian Cho, Paul Walker**

Collaborative Assessment for Amphibian Conservation

Zoe Goodwin and Paul Walker, Wildtracks

Half of all amphibian species around the world are now listed Near Threatened, Vulnerable, Endangered, or Critically Endangered. Most of declines in amphibian populations of pristine areas in Central America have occurred at elevations of over 500 m, but little research work has been conducted in these areas of Belize.

In July 2006 Wildtracks and Kris Kaiser (University of Miami), in collaboration with the Belize Audubon Society and the Institute of Archaeology, conducted a pilot assessment of the status of Belize's endangered amphibian species. Field surveys were conducted within three areas of Cockscomb Basin Wildlife Sanctuary (Cabbage Haul Ridge, Outlier, and along portions of the Cockscomb Range), as well as in the Caracol Archaeological Reserve.

The study was also used as a training opportunity to build the field recognition and monitoring capacity of Wildtracks and BAS staff members. Norman Mora, Ivis Chan, and CBWS Warden Bonifacio Cal, along with Emmerson Garcia (University of Belize), and two porters from Maya Center were amongst the ten-person team who conducted the work on the ridges beneath the towering peak of The Molar in the Cockscomb Range. A total of 14 species of amphibian and 17 reptile species were observed in the Cockscomb surveys (*table, right*)

One of the amphibians is almost certainly new to science, possibly of the *Craugastor* genus. Two specimens were preserved and sent to a regional amphibian taxonomist for validation and description. *Craugastor laticeps* (formerly *Eleutherodactylus laticeps*) is a new record for Cockscomb. Of the reptiles observed in Cockscomb, the diminutive *Adelphicos quadrivirgatus* snake was a new record for the Sanctuary.

Amphibians and Reptiles in Cockscomb Basin Wildlife Sanctuary

Amphibians found in CBWS	
Bolitoglossa rufescens	Rufescent salamander
Bufo campbelli	Rainforest toad
Bufo marinus	Marine toad
Bufo valliceps	Gulf Coast toad
Craugastor chac	Maya rain frog
Craugastor laticeps	Broad-headed Rainfrog
Craugastor loki	Lowland rain frog
Eleutherodactylus leprus	Leprus chirping rain frog
Eleutherodactylus sp.	New species ?
Agalychnis callidryas	Red-eyed treefrog
Smilisca baudinii	Mexican treefrog
Hyalinobatrachium fleischmanni	Glass frog
Rana juliani	Maya Mountain frog
Rana vaillanti	Rainforest frog
Reptiles found in CBWS	
Anolis capito	Big-headed anole
Anolis lemurinus	Ghost Anole
Anolis tropidonotus	Greater scaly anole
Anolis uniformis	Lesser scaly anole
Basiliscus vittatus	Striped basilisk
Corytophanes cristatus	Old man lizard
Eumeces sumichrasti	Sumichrast's skink
Mabuya unimarginata	Bronze skink
Sphenomorphus cherriei	Cherrie's skink
Ameiva festiva	Central American whiptail
Ameiva undulata	Barred whiptail
Adelphicos quadrivirgatus	Zacatera snake
Imantodes cenchoa	Blunt-headed tree snake
Lampropeltis triangulum	Tropical kingsnake
Leptodeira septentrionalis	Cat-eyed snake
Ninia sebae	Red coffee snake
Micrurus diastema	Coral snake

PLANT INVENTORY IN CBWS

STEVEN W. BREWER, UNIVERSITY OF NORTH CAROLINA

The central and western CBWS contains the only relatively undisturbed tropical rainforest on "wet" granite in Belize, and perhaps anywhere in northern Central America, in contrast to the highly disturbed forests of low rainfall ("dry" granite) located in the Mountain Pine Ridge of the northwest side of the Maya Mountains. During the latter part of November 2006, for approximately three weeks Marcelo Pau, Juan Chub, Ricardo Chun, Timothy Sho, Adolfo Pop and I conducted a preliminary forest census in the north-central CBWS. The objectives of this census were to sample tree diversity, physical structure, and species composition of intact, upland forests on granite substrate; compare the forest of the CBWS with forests on other geological substrata elsewhere in the Maya Mountains; and test the need and feasibility of further plant inventory and plant ecological studies in the CBWS.

On each of two ridge systems perpendicular to the Cockscomb Range, we surveryed three topographic positions: 1) lower slope, 2) upper slope, and 3) ridge. In each of those six areas we laid out 2-meter by 500-meter belt plots (transects) in which we censused trees. We measured diameter at breast height (DBH) and identified, to the species, all individuals that were greater than 2 cm DBH. Herbarium specimens were collected for all species that were not readily identifiable in the field. A few herbarium specimens were also collected for woody plants, plus tree ferns and palms. Herbarium specimens are to be shipped to Belize for deposit in the Belize National Herbarium (BRH).

In physical structure, the CBWS has comparable, if somewhat lower, tree density than forests on volcanics. However, forests on granite and volcanics are significantly denser than those on limestone.

In species composition, the CBWS transects represent distinctly different forests than those found on volcanics and limestone. Compared to forests on limestone and volcanics, the CBWS transects were comparatively rich in species of the *Lauraceae*, *Moraceae*, and *Sapotaceae* families. However, the *Rubiaceae* family was comparatively poorly represented in the CBWS transects compared to forests on other substrata. The CBWS forests appear to have unusually high fern (*Pteridophyta*) diversity for the country, including all of the tree fern species known to Belize.

Some notable species encountered include a new record for Belize *Drypetes guatemalensis* (*Euphorbiaceae*), a very rare species throughout its known range. Also discovered—unfortunately in sterile condition—is what appears to be a possible new species to science in the Zygia complex of mimosoid legumes (*Fabaceae*), however fertile specimens will be required for determination.

Alcorneopsis floribunda (*Euphorbiaceae*), and *Chamaedorea costaricana* (*Arecaceae*), rare and recently discovered in the Bladen Nature Reserve as new records for Belize, were also found in the CBWS.

Roupala glaberrima (*Proteaceae*) was commonly observed at elevations above 400 m. Prior to collection of this species on Doyle's Delight in 2004 by S. Brewer and S. Bridgewater, this species was known only from one specimen collected by William A. Schipp in 1934 and only determined (2002) by G.T. Prance.

Talisia macrophylla (*Sapindaceae*) and *Alsophila salvinii* (*Cyatheaceae*) were collected in CBWS, making these collections only the second for Belize.

Other rare or uncommon species for Belize found in the CBWS include *Rinorea deflexiflora* (*Violaceae*), *Virola multiflora* (*Myristicaceae*), and *Sphaeropteris horrida* (*Cyatheaceae*).

Table 5.6 Species in the Studied Transects at 500 meters

Species Name	Family*
Saurauia yasicae Loes.	ACT
Metopium brownei (Jacq.) Urb.	ANA
Spondias radlkoferi	ANA
Annona liebmanii	ANN
Cymbopetalum mayanum Lundell	ANN
Guatteria amplifolia	ANN
Mosannona depressa (Baill.) Chatrou	ANN
Xylopia frutescens Aubl.	ANN
Aspidosperma megalocarpon Müll. Arg.	APO
Aspidosperma spruceanum Benth. ex Müll. Arg.	APO
Stemmadenia donnell-smithii (Rose) Woodson	APO
Desmopsis schippii	AQU
Ilex belizensis	AQU
Bactris mexicana Mart.	ARE
Chamaedorea ernesti-augustii H. Wendl.	ARE
Chamaedorea neurochlamys	ARE
Chamaedorea tepejilote Liebm.	ARE
Desmoncus orthacanthos Mart.	ARE
Euterpe precatoria Mart. var. longevaginata (Mart.) Henderson	ARE
Geonoma deversa (Poit.) Kunth	ARE
Geonoma interrupta (Ruiz & Pav.) Mart. var. interrupta	ARE
Synechanthus fibrosus (H. Wendl.) H. Wendl.	ARE
Dendropanax arboreus (L.) Decne. & Planch.	ARL
Eupatorium /Zexmenia or segregate	AST
Amphitecna breedlovei A. H. Gentry	BIG
Quararibea funebris (La Llave) Vischer	BOM
Cordia bicolor A. DC.	BOR
Bursera simaruba (L.) Sarg.	BUR
Protium confusum (Rose) Pittier	BUR
Forchammeria trifoliata	CAP
Crossopetalum parviflorum (Hemsl.) Lundell	CEL
Licania hypoleuca Benth.	CHB
Licania sparsipilis S. F. Blake	CHB
Calophyllum brasiliense Cambess.	CLU
Garcinia intermedia (Pittier) Hammel	CLU
Marila laxiflora	CLU
Symphonia globulifera	CLU
Terminalia amazonia (J. Gmel.) Exell	CMB
Jacaratia dolichaula (Donn. Sm.) Woodson	CRC
Sloanea meianthera Donn. Sm.	ELC
Sloanea tuerckheimii Donn. Smith	ELC
Croton arboreus Millsp.	EUP
Croton billbergianus ssp. pyramidalis	EUP
Drypetes guatemalensis	EUP
Hieronyma alchorneoides Allemao	EUP
Sapium laurifolium	EUP
Abarema idiopoda (S. F. Blake) Barneby & J. W. Grimes	FAB
Cojoba graciliflora (Blake) Britt. & Rose	FAB
Dialium guianense (Aubl.) Sandwith Erythrina sp. 2; 3420	FAB
Inga belizensis	FAB
Inga davidsei M. Sousa	FAB
Lonchocarpus aff. purpureus; 3375	FAB
Lonchocarpus heptaphyllus (Poir.) DC.	FAB
Pterocarpus rohrii Vahl	FAB
Schizolobium parahyba (Vell.) S.F. Blake	FAB
Swartzia simplex var. ochnacea (DC.) R.S. Cowan	FAB
Casearia arborea (Rich.) Urb.	FLC
Casearia sylvestris	FLC
Casearia tremula (Griseb.) Griseb. ex C. Wright	FLC
Lunania mexicana Brandegee	FLC
Macrohasseltia macroterantha (Standl. & L. O. Williams) L. O. Williams	FLC
Pleuranthodendron lindenii (Turcz.) Sleumer	FLC
Calatola costaricensis Standl. (check C. laevigata)	ICC
Nectandra hihua	LAU
Nectandra longicaudata (Lundell) C. K. Allen	LAU
Nectandra lundellii C. K. Allen	LAU
Ocotea cernua (Nees) Mez	LAU
Ocotea helicterifolia (Meisn.) Hemsl.	LAU
Ocotea leucoxylon (Sw.) Laness.	LAU
Lacistema aggregatum (Berg.) Rusby	LCS
Guarea glabra Vahl	MEL
Swietenia macrophylla King	MEL
Bunchosia lindeniana aff.	MLP
Micona elata	MLS
Miconia impetiolaris (Sw.) D. Don ex DC.	MLS
Mouriri exilis Gleason	MLS
Brosimum guianense (Aubl.) Huber	MOR
Brosimum lactescens (S. Moore) C. C. Berg	MOR
Ficus guajavoides Lundell	MOR
Ficus insipida Willd.	MOR
Pourouma bicolor Mart.	MOR
Pseudolmedia spuria (Sw.) Griseb.	MOR
Trophis mexicana (Liebm.) Bureau	MOR
Myrsinaceae 3310	MRS
Myrtaceae sp. 1	MRT
Myrtaceae sp. 2	MRT
Plinia peroblata	MRT
Compsoneura mexicana (Hemsl.) Janovec	MYS
Virola koschnyi Warb.	MYS
Virola multiflora	MYS
Neea psychotrioides Donn. Sm.	NYC
Podocarpus guatemalensis	PDC
Picramnia antidesma	PIC
Piper psilorhachis	PIP
Coccoloba belizensis Standl.	PLG
Roupala montana	PRT
Cyathea myosuroides	PTE
Cassipourea elliptica (Sw.) Poir	RHZ
Faramea belizensis	RUB
Hamelia sp.	RUB
Posoqueria latifolia (Rudge) Roem. & Schult.	RUB
Psychotria chiapensis Standl.	RUB
Randia matudae	RUB
Esenbeckia pentaphylla	RUT
Zanthoxylum ekmanii (Urb.) Alain	RUT
Zanthoxylum riedelianum Engl.	RUT
Cupania macrophylla aff.	SAP
Matayba apetala Radlk.	SAP
Simarouba simaroubensis	SIM
Mollinedia guatemalensis Perkins	SIP
Manilkara chicle (Pittier) Gilly	SPT
Manilkara staminodella Gilly	SPT
Pouteria belizensis	SPT
Pouteria campechiana (Kunth) Baehni	SPT
Pouteria durlandii Standl. (Baehni)	SPT
Pouteria izabalensis (Standl.) Baehni	SPT
Pouteria sapota	SPT
Sideroxylon sp. cf.	SPT
Sideroxylon floribundum ssp. belizense (Lundell) T.D. Penn.	SPT
Symplocarpon purpusii	THE
Ternstroemia tepezapote	THE
Luehea speciosa Willd.	TIL
Mortoniodendron sp. cf.	TIL
Rinorea guatemalensis (S. Watson) Bartlett	VIO
Rinorea hummelii Sprague	VIO
Vochysia guatemalensis	VOC

*Family Acronyms from Weber <unco.edu/nbs/biology/environment.herbarium/acronyms>

Xaté Project

Ivis Chan

The leaves of the understory xaté palm *Chamaedorea ernesti augustii,* commonly known as the fishtail palm, are economically valuable and are harvested to supply foliage for foreign floral markets. Much of the extraction of this non-timber forest product is conducted illegally in Belize. In addition to the impact of leaf harvesting on *Chamaedorea* population health, this activity places other pressures on natural resources including over-hunting of game species, such as peccaries, and hunting of flagship species such as scarlet macaws. Incursions now extend into nationally protected areas, but few data exist on the associated impacts.

(*above*) TMNR Warden Albart Liconia and Protected Areas Manager Dominique Lizama measuring xaté (*right*) Xaté (*Chamaedorea ernesti augustii*)

This study presents a preliminary assessment of the extraction of xaté within two protected areas co-managed by the Forest Department and the Belize Audubon Society. The aim was to identify the xaté species occurring within them and to establish baseline data on illegal harvesting. Over a six-month period more than thirty plots were surveyed within Tapir Mountain Nature Reserve and St. Herman's Blue Hole National Park. For each plot the species of xaté palm encountered was recorded as well as data regarding abundance, height, number of leaves present and leaf length. For *C. ernesti augustii,* the number of cut leaves and number of leaves considered to be of commercial value were also recorded.

Preliminary results indicate that at least four species of xaté occur in these parks: *C. ernesti augustii*, *C. tepejilote*, *C. neurochlamys*, and *C. oblongata*. The most abundant species at St. Herman's Blue Hole was *C. tepejilote* followed by *C. ernesti augustii*, while *C. ernesti augustii* and *C. neurochlamys* were most abundant at Tapir Mountain. Almost 90 percent of all plots contained individuals of *C. ernesti augustii* showing evidence of leaf harvesting.

These results highlight the countrywide impact of the xaté harvesting industry as they provide evidence that the pressure of illegal harvesting now extends further than the western reaches of the Maya mountains and the border of Belize with Guatemala.

For this project Research Manager Ivis Chan was assisted by TMNR wardens Larry Santos and Albart Liconia; Peace Corps Volunteer Scott Schmidt and Protected Areas Manager, Dominique Lizama, SHBHNP Warden Israel Manzanero Jr., and hired helper Celso Sho.

Marine Forum

On the 12th December 2002, the Belize Audubon Society with the support of United Nations Development Programme, Global Environmental Facility, COMPACT, Programme for Belize, Wildlife Conservation Society and the United Nations Foundation hosted the first ever National Marine Forum. The intent was to showcase the works being carried out in Belize's marine protected areas and to highlight major issues affecting our marine ecosystems and resources.

Marine Biologists from the various protected areas as well as distinguished scientists working in the Belizean marine environment presented the findings of their research and monitoring activities. The forum also provided the opportunities to hear the views and concerns of other marine stakeholders, such as, the fishermen cooperatives, the tourism initiatives and the governmental sectoral and departmental managers.

James Azueta, Janet Gibson, and Melanie McField on a panel at the Marine Forum

Former BAS Marine Protected Areas Manager Julianne Robinson with Dr. Ed Boles of the University of Belize

BAS Research Coordinator Roberto Pott presenting spawning aggregation studies at the Marine Forum.

BAS Marine Biologist Roger Arana presenting conch and lobster surveys at the Marine Forum.

Limits of Acceptable Change for the Blue Hole and Half Moon Caye Natural Monuments

Marianna Dominguez

According to the 2007 HMCNM and BHNM Management Plan, a total of 138 species of fish, 48 species of scleractinian corals (stony corals), 46 species of non-coral invertebrates, 128 species of resident and migratory birds and 10 reptile species have been recorded at Half Moon Caye and Blue Hole Natural Monuments. These areas provide protection for fifteen species of concern under the IUCN Redlist programme (rated as Critically endangered, Endangered or Vulnerable), including the white-phase red-footed boobies, the hawksbill, loggerhead and green turtles, and the American crocodile.

The methodology of Limits of Acceptable Change (LAC) admits that anthropogenic impacts within protected areas are inevitable, but establishes limits for the acceptable degree of change. The basic concept includes a determination of the conditions that the area should have and of the impacts that may affect these conditions; the establishment of indicators and standards for those indicators that should not be surpassed; and a monitoring program to continuously evaluate the impacts caused by human activities.

To determine the current status of the coral reef communities, different habitat types found within the protected areas were evaluated: fore reef, patch reef, and back reef habitats were assessed in Half Moon Caye Natural Monument, and the coral rim habitat for the Blue Hole Natural Monument. Three randomly chosen sites were assessed for each type of habitat on recreational areas, and one site was evaluated on the pristine zone, for a total of 13 sites. Five transects were deployed in each site and the Line Point Intercept method was used to determine percentage cover of sessile organisms. A characterization of the first 50 coral colonies found in these transects was done in order to determine the community composition, size distribution, extent of bleaching, frequency of coral diseases and mortality.

The mean live benthic cover refers to the entire benthic reef community, which includes hard corals, algae, gorgonians, sponges, and encrusting coralline algae. When grouping all of these components together the mean live benthic cover for the total area studied is 63.87% (S.D. 11.30%) (*see table below*).

Mean live benthic cover in HMCNM and BHNM	
Habitat	Mean Live Cover
Blue Hole rim	75.56%
Patch reef HMC recreational area	68.83%
Back reef HMC recreational area	57.94%
Fore reef HMC recreational area	69.83%
Back reef HMC conservation area	47.17%
Total area	**63.87%**

The effective carrying capacity, the maximum amount of visitors that should be allowed in the protected area, was calculated to be as follows.

Carrying Capacity of HMCNM and BHNM		
Protected Area	Visitors/day	Visitors/year
HMC Trail	52	18,980
BHNM Divers	15	5,475
HMC Divers	25	9,125

In comparison, the number of divers in 2008, were 6,425 in HMCNM and 6,174 in BHNM. The carrying capacity could be increased by 76 percent by increasing the management personnel, ideally adding two more wardens, another park director, another coxswain, an assistant to the manager, a liason officer, and an educator.

Marine Research

Julianne Robinson, Jamie Lee Chanona, Sergio Hoare, and Charles Acosta

Assessment of Diving Impact

Two sites were established in 1999 for assessing the impact of diving on the reef. Video transect monitoring was used.

Conch and Lobster Surveys

Ten sites were surveyed, three patch reefs inside the HMCNM and three outside, two flats areas inside the HMCNM and two outside. Three species of conch and five species of lobster were counted and measured. The results show that there is a core area within the Half Moon Caye Natural Monument that is crucial for the reproduction cycle of the Queen Conch.

Habitat Map

A habitat map for the Blue Hole and Half Moon Caye Natural Monuments were drawn and digitzied using aerial photographs. Ground control points were collected and ground truthing showed no changes in sizes or distribution of patch reefs.

Monitoring Whale Sharks

A receiver was installed at Half Moon Caye to track radio tagged whale sharks. One whale shark that was tagged in southern Belize in April, 2001, passed Half Moon Caye on May 17th.

Mesoamerican Caribbean Reef Program

BAS worked with World Wildlife fund's scientific reef survey of the entire Belize Barrier Reef System initiated in 2001.

Carlos Ayala

Conch (*Strombus gigas*)

Sandy Webb

Whale shark (*Rhincodon typus*)

Spawning Aggregation

Roberto Pott

Beginning in 2000, spawning aggregation banks were assessed at Lighthouse Reef in a cooperative project between BAS, Green Reef, The Nature Conservancy, and Toledo Institute for Development and the Environment, called the Belize Spawning Aggregation Working Group (SPAG). Initially two banks were found, one within the Half Moon Caye Natural Monument, and the other unprotected.

The first Spawning Aggregation Monitoring Training was held from March 23rd to April 7th, 2002. In addition to the organizations listed above, the Fisheries Department, Coastal Zone Management Authority and Institute, and the University of Belize also participated. Attending were students, fishermen, scientists and technicians from several countries in the Caribbean and from Hong Kong. Data was collected at several sites and satellite imagery was used to identify possible additonal spawning aggregation sites.

Research by SPAG resulted in the legal protection of eleven spawning aggregation sites and closure of the Nassau Grouper fishery during their spawning season from December to March.

By 2008 SPAG reported such serious depletion of the Nassau Grouper population in Belize that a nation-wide moratorium on the take of Nassau Groupers was recommended. The law now requires that fillets of fish be sold with a patch of skin at least one inch square to make sure that they are not Nassau Groupers. Since most Nassau Groupers are caught by spear fishing, a ban on spear fishing was also recommended by SPAG.

Doug Perrine

Nassau Grouper spawning aggregation

BAS Staff Research Results

In 2008-2009, the marine staff noted that the HMCNM has the largest most stable Nassau Grouper spawning site in Belize; conch species is a viable healthy population with 2000 per hectare; the sea grass is healthy, and the three species of turtles continue to use Half Moon as a nesting site. The corals saw a decrease in coral bleaching from 10%-2%.

Monitoring at Half Moon Caye and Blue Hole Natural Monuments

Celso Cawich and Losita Lee

In 2009, BAS marine monitoring included seagrass monitoring using the Sea Grass Net Monitoring Protocol, Nassau Grouper Aggregation Site monitoring, Conch and Lobster population density surveys, coral bleaching monitoring using the Weighted Bar Swimming Transect Method, and Sea Turtle Monitoring using the IUCN Sea Turtle Specialist Group Monitoring Protocol. Results of this ongoing monitoring are as follows:

- Anthropogenic impacts were 16% at sea grass sites.
- The Nassau Grouper population at the aggregation site 3 km southeast of Sand Bore Caye was 2,050, a 64% increase from the 2008 count.
- The conch population density for HMCNM was estimated at 2,087 per hectare with 92% juveniles; 85% of the conch was in protected areas.
- Three species of marine turtles—hawksbill, loggerhead, and green—nest at Half Moon Caye; hatchling success was estimated at 88%.
- Coral bleaching at the Blue Hole was highest in October at 42%.

These monitoring data assisted the protected areas management in prioritizing resource allocation and patrolling, designation of turtle nesting grounds, conservation of key species, and promotion of tourism best practices for the two protected areas.

These results of BAS Staff biodiversity monitoring were presented at the University of Belize's Natural Resources Management Department and at the XIII Congress of the Mesoamerican Society for Biology and Conservation in October, 2009.

Sea Grass Monitoring 2006-2010

Amador Pott and Eliceo J. Cobb

BAS Staff monitored two seagrass sites, one located on the northern part ofthe island in HMCNM at a depth of 6.3 feet. At BHNM the study site is in the northeastern entrance to the Blue Hole at a depth of 12.5 feet.

Two species of sea grass have been recorded at the BHNM, turtle grass (*Thalassia testudinum*) and manatee grass (*Syringodium filiforme*). Brown algae also occurs, which is an indication of stress, such as heavy boat traffic in the area. In another indication of stress, the brown algae is larger at the BHNM sites than that found in the HMCNM monitoring site. Three transects were studied at each site, labeled A, B, and C.

In addition to turtle grass and manatee grass, shoal grass (*Halodule wrightii*) is recorded at the HMCNM monitoring site. Manatee grass was observed flowering in January and May of 2010. Turtle grass was observed flowering in August 2010. No flowering has been observed for shoal grass.

Twelve monitoring sessions were conducted at each site, three times per year for four years. All transects showed an increase in sea grass coverage over four years (*Table below*).

	A	B	C
HMCNM	8%	25%	27%
BHNM	37%	43%	30%

The increase has been greater in BHNM than in the HMCNM, which may be because the HMCNM site is at a shallower depth.

Lobster and Conch Monitoring at Lighthouse Reef

Amador Pott and Eliceo J. Cobb

Lobster and conch monitoring took place at 23 sites at Lighthouse Reef, seven sites within HMCNM, four sites within the BHNM, and twelve outside protected areas.

The lobster underwater survey was conducted in 2009 and 2010 using the AGRRA protocol in March after the closing of the lobster season and in June just before the opening of the season.

The results of lobster monitoring show that 45 percent of lobster were found in HMCNM, 27 percent outside the protected areas and 25 percent in BHNM. In 2009 adults formed the largest proporation on the population, while in 2010 most lobsters found were juveniles.

Conch monitoring was conducted just after the closing of the conch season in July and immediately before opening of the season in the end of September.

Most of the conchs were found HMCNM and two sites outside protected areas. The population of conchs were distributed as 70 percent juveniles and 30 percent adults, half of which were reproducers.

The results indicate that HMCNM serves as a healthy nursery habitat providing the proper environment for conch growth and development.

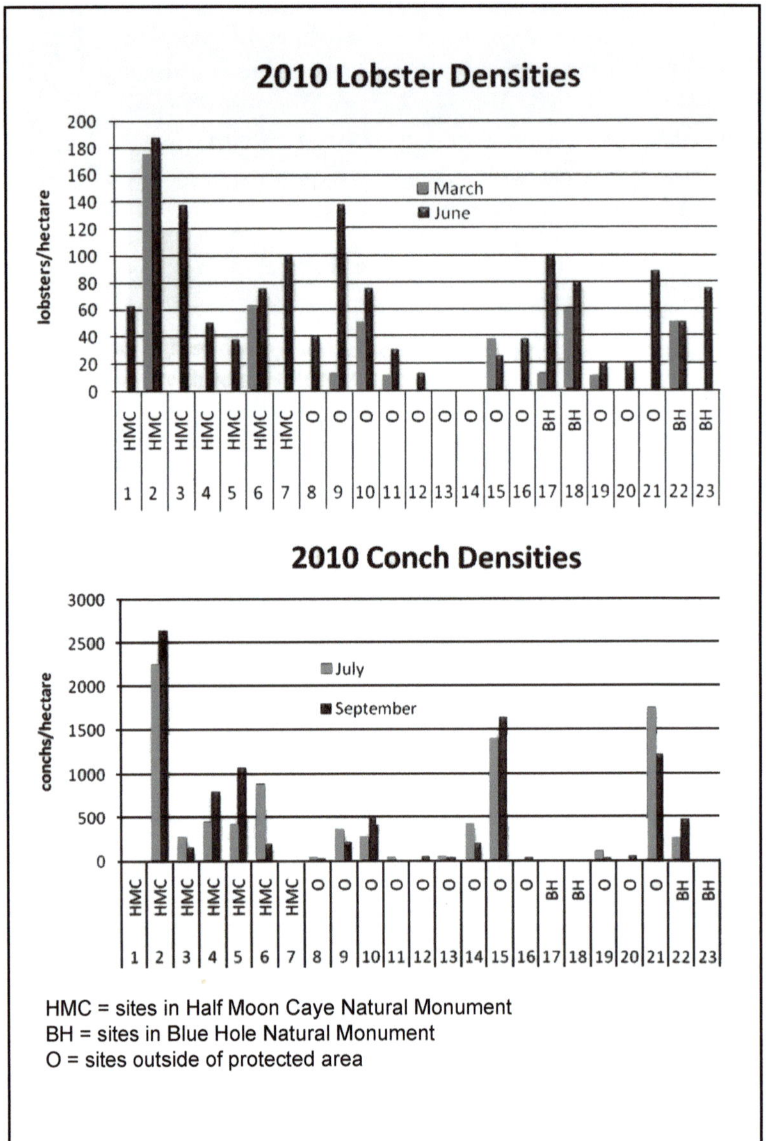

Sea Turtle Monitoring

BAS Staff monitors turtle nesting on Half Moon Caye. They check each instance of turtle tracks where turtles have crawled onto the beach. Some resulted in nests and others were merely turtle crawls. In 2008 there were thirteen instances of turtle tracks, 11 in July and two in September. There were four nests altogether, two in July and in September. In 2010, there were eight turtle tracks in June, 10 in July, five in August, and three in September.

Turtle Monitoring				
Year	Jun	Jul	Aug	Sep
2008		2/11*		2/2
2010	3/8	4/10	2/5	1/3

* # of nests/# of turtle tracks

In 2008 two of the nests were hawksbill turtles; one was a green turtle nest; and one was a loggerhead turtle nest. In 2009 there was only one nest and it was a loggerhead nest. In 2010 five nests were hawksbill, one was a green, and one loggerhead turtle nest. The species was not identified in three of the nests. Two of those nests washed away in weather events.

The 2010 clutch sizes ranged from 90 to 142 eggs. The green turtle nests had the highest hatchling success rate with 97 percent hatchlings. The hawksbill had 91 percent hatchling survival. Due to human interference, the first hawksbill nest only had 18 percent hatchling success. The second hawksbill nest had 32 percent hatchling success, but the bottom of the nest was wet and some embryos didn't develop normally. The third hawksbill nest had no surviving hatchlings because the whole nest was submerged under water.

Lionfish Eradication

BAS has a programme for the eradication of red lionfish (*Pterois volitans*) and devil fire fish (*P. miles*). Both species are native of the Indian/Pacific Ocean that have invaded the western Atlantic Ocean and Caribbean Sea. They constitute an unpredictable threat for several reasons. They have venomous spines that are a threat to fishermen, divers, snorkelers, and swimmers. They are aggressive carnivorous fish that could deplete native reef fish, both as a predator and as a competitor for food.

The Lionfish Eradication Programme involved a public education programme in which everyone was encouraged to report and, if possible, capture any lionfish found in Belizean waters. BAS staff captured any lionfish encountered during monitoring work and studied those turned in by dive operators. Basic data were collected from each specimen: habitat where captured, depth at which it was captured, sizes, colour, presence of eye stalks, and sex.

Sixty-five lionfish were captured in 2010 from different habitats in Lighthouse Reef Atoll. The habitat where the largest percentage of the population were captured was in patch reef (43%), second was back reef (37%), then in the wall (17%) and the lowest in the wreck (fore) reef. These percentages may reflect the time spent by divers or researchers in these habitats rather than the actual distribution of the lionfish. Lionfish prefer deeper areas where water current is minimized. Colours of the fish captured were orange (52%), black (40%) or red (8%). All had eye stalks.

Although the eradication programme has resulted in a number of captures, lionfish continue to be found in Belizean waters.

Lighthouse Reef Marine Research—2010

Rachel T. Graham, James P. Lewis, Adrian C. Gleiss, Kylie Scales, Sophie Thompson

The primary objective of the research sponsored by the Ocean Giants Program of the Wildlife Conservation Society in 2010 was to continue the implementation of projects founded on the pilot assessment activities in 2006 and project implementation in 2007 to gather data on top predator abundance and behaviour at our study sites in Southern Belize and Lighthouse Reef Atoll. Our specific activities in 2010 were to:

- Monitor relative abundance of elasmobranchs at Lighthouse Reef Atoll and continue an acoustic tracking network to monitor their movements in relation to the atoll's habitats and the Blue Hole,
- Assess hawksbill turtle use of habitat through active tracking at Lighthouse Reef Atoll,
- Monitor the catch rate and long distance movements by goliath grouper, and
- Monitor the national acoustic network to track sharks, whale sharks and other marine megafauna.

We finalised field work on the goliath grouper project with the notable additions of recaptures of tagged fish in Mexico, Turneffe Atoll and Glover's Reef Atoll indicating that these juveniles are willing to cross deep waters to new grounds and possibly new spawning sites.

We supported SACD and Wildtracks in a rapid assessment of elasmofauna in the lagoon system of northern Belize in the hopes of finding evidence of smalltooth sawfish, but were unfortunately disappointed—further proof that this species is ecologically extinct in Belize.

Caribbean whiptail stingray (*Himantura schmardae*) was a first for Lighthouse Reef Atoll.

The Lighthouse elasmobranch research revealed the existence of a ray *Himantura schmardae* that had not previously been documented at the atoll and which provided us with new records of size at maturity for both males and females. There was evidence of gravid females, males and juveniles, there is scope for further study on this understudied species to establish distribution, habitat use and the identification of locations for breeding and parturition.

The *Dasyatidae* family of stingrays, along with many other elasmobranchs, are becoming an increasingly important fisheries resource in Belize, exacerbated in part by the decline of many shark species. As exploitation continues to occur, vital information is still lacking. An understanding of abundance, movement, and habitat utilisation of ray species of conservation concern is a prerequisite to the management plans in Belize involving reserves such as protected areas and no-take zones to protect species from overexploitation. Little is known about the ray species in Belize, particularly at Lighthouse Reef Atoll.

Five *D. americana* and three *Himantura schmardae* (Caribbean whiptail stingray) were tagged with active tracking transmitters and tracked for sixteen days. By tracking rays actively it is possible to establish which habitats are important for the species. Home ranges can also be established to identify the area the animal regularly travels for food or shelter. In addition, forty-nine in-water transects were completed across the atoll, to establish the presence and distribution of rays.

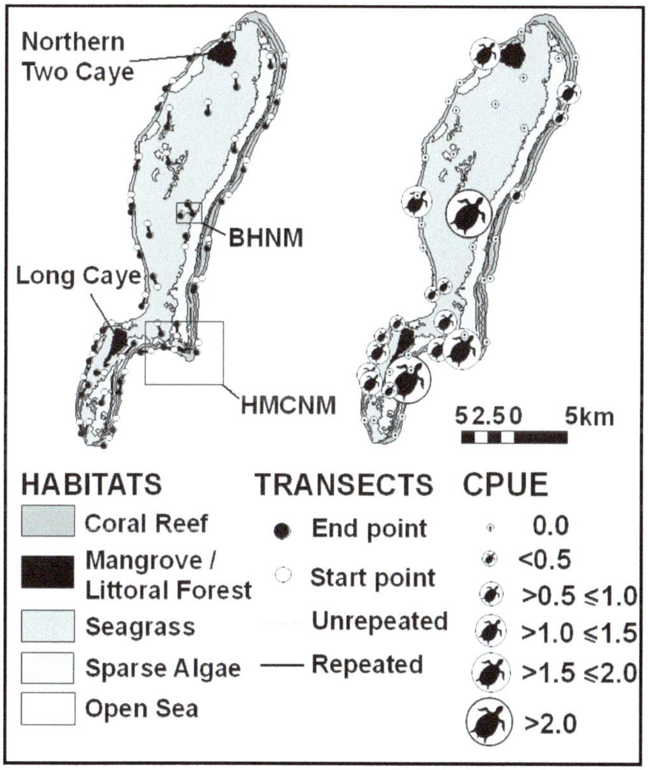

Randomised transect locations (*left*) and turtle abundance as CPUE (*right*) at Lighthouse Reef Atoll, Belize.

The first great hammerhead to be tagged. The tracking indicated the hammerhead shark is highly vulnerable to night-set subsurface nets.

Expansion of the turtle tracking research initiated in 2009 proved that hawksbills at Lighthouse possess the largest known home ranges for juveniles of this species. Turtles are more numerous in protected areas (BHNM, HMCNM) and in the coral reef habitat than in the lagoon (*left*).

We used tagged sharks to look at the cost of carnivory in sharks. Capturing large elusive prey requires a lot of power. Results show that the Caribbean reef shark used quick bursts of speed and movement in the water column occupying similar depths day and night while the great hammerhead sharks have a diurnal preferences with highly active night time hunting taking place in shallow, likely lagoonal waters and daytime more tranquil foraging in deeper forereef waters. We provided support to the Fisheries Department in their bid to propose regulations that require that fins remain on landed sharks.

We implemented a national survey on public perceptions towards sharks. We will be disseminating this information to the public and to our peers, which we hope will bolster management and conservation efforts. We look forward to the continuation of our partnerships with Government, academia, the NGO community, the private sector and fishers and guides.

We have continued to raise awareness for sharks through international means by collaborating with individuals such as National Geographic's premier photographer Brian Skerry on a story on the Mesoamerican Barrier Reef that is expected to highlight Belize and sharks. The episode on whale sharks and spawning fish on BBC's series "Life" released in early 2010 was guided by WCS. WCS proposed and guided the National Geographic's episode on whale shark migrations along the Mesoamerican Barrier Reef in the series entitled "Great Migrations," which was released in November to a broad international audience.

Recommendations

1) Continue to integrate sharks and rays and goliath grouper in all protected area management plans that do not currently include or address these species.

2) Ban gillnets and longlines in Belize. This will support the sports fish and herbivore protection, as well as decrease the impact of transboundary fishing on Belize's sharks and rays.

3) Ban the fishing of sharks and goliath grouper in Belize to help increase reef resilience, strengthen overnight marine tourism thus boosting local economies, and very importantly as our best option for biologically controlling lionfish (*Pterois volitans*), a prolific invasive species that has exploded in numbers throughout Belize's coasts and reef.

Summary of Research for Biodiversity and Conservation

From the beginning, BAS has been involved with research. The first circle (Belize City) for a Christmas Bird Count (CBC) was established the first year of BAS existence. There are now five CBC circles and Important Bird Areas designated. Jabiru, Red-footed booby, and other bird counts in the Spring and Fall migrations all add to the forty years of data on birds at BAS.

Research in Crooked Tree Wildlife Sanctuary includes water quality assessment and investigations into the invasion of the exotic *Tilapia* fish species. Collaboration with the El Colegio de la Frontera Sur in Chetumal, Mexic manatee research began when a tagged male manatee called "Manchas" made his way into Crooked Tree Lagoon.

The Cockscomb Basin Wildlife Sanctuary (CBWS) was first declared after research on jaguars was carried out by Alan Rabinowitz. Other research has been done in the CBWS on plants, butterflies, damselflies, fish, reptiles, amphibians, bats, birds, and mammals. A successful translocation of Black Howler Monkeys (*Alouatta pigra*) from the Community Baboon Sanctuary on the Belize River restored that species to the Cockscomb Basin.

An assessment of the economically valuable xate palm showed the impact of harvesting in all of the BAS-managed Cayo Parks, as far east as St. Herman's Blue Hole National Park.

For many years, BAS staff at Half Moon Caye and Blue Hole National Monuments have monitored seagrass, coral, finfish, lobster, conch, stingrays, sharks, and sea turtles in Lighthouse Reef Atoll.

Biodiversity monitoring is an integral part of the management of BAS-managed protected areas.

Chapter Six

Awards

James A. Waight Conservation Award

First Ceremony - 1987

The first James A. Waight Award ceremony was held at the Chateau Caribbean on Friday evening, April 10, 1987. Vice-president Janet Gibson introduced the President of the Belize Audubon Society, Dr. Victor Gonzalez, who gave the history of the award, named the James A. Waight Conservation Award in honour of the Society's first president. In expressing his thanks and appreciation of the honour, Mr. Waight paid tribute to Mrs. Dora Weyer, who founded the Belize Audubon Society, and to the founder members who were still actively participating in the work of the Society.

Mr. Waight then presented the first James A. Waight Award to Squadron Leader Richard Charles Atkinson, who has provided invaluable assistance to the Belize Audubon Society in demarcating the boundaries of the Half Moon Caye Natural Monument and promoting the Reef Preservation Fund. In turn, Squadron Leader Atkinson presented a Royal Air Force wall shield to the Belize Audubon Society.

BAS Archives

(*above*) At the first James A. Waight Award Ceremony James Waight (*left*) presents award to Squadron Leader Richard Charles Atkinson. (*right*) Wall Shield given to BAS by the Royal Air Force

1988—W. Ford Young

The second James A. Waight Award was given to Ford Young, one of the BAS founding members. He contributed a lot of time and effort in advancing the conservation of the natural heritage of Belize. He served for many years as the Vice President of the Society. He was generally in charge of the monthly field trips that the Society supported. Mr. Young coordinated the Christmas Bird Counts and assisted with aerial surveys in keeping track of the Jabiru nesting habits and sites in the country of Belize. These are but a few of the many activities that Mr. Young was involved in during his years on the Board of Directors of the BAS. The Society presented him with the James A. Waight Conservation Award in gratitude for his service to the Society.

The Community Award was presented to the Community Baboon Sanctuary. In 1985 nearly all the landowners in an eighteen-square-mile area near the Belize River voluntarily signed pledges that committed them to protect the habitat of the black howler monkey in that region. Their commitment made possible the establishment of the Community Baboon Sanctuary, a grassroots conservation programme. The communities involved in this project include: Bermudian Landing, Big Falls, Double Head Cabbage, Flowers Bank, Isabella Bank, St. Paul's and Willows Bank. BAS played an active role in the early management of the sanctuary and worked in close cooperation with the Sanctuary Manager, Mr. Fallett Young. Dr. Robert Horwich was a key player in the establishment of the Sanctuary.

1989—Dora Weyer

Dora Weyer was recognized for her services to the Society. She was a founding member of the BAS and served on its Board of Directors until 1987. She was an active member of the Board and was instrumental in bringing about the establishment of the Crooked Tree Wildlife Sanctuary. Numerous other projects have been made possible through her untiring efforts.

At the same awards ceremony, Walter (Micky) Craig, Executive Director from 1985 to 1989, was presented with a zericote carving in recognition of his services. Esso Standard Oil, S.A, received a plaque in appreciation of their donation.

Mexican Black Howler Monkey (*Alouatta pigra*)

1990—Dr. Victor Gonzalez and Alice M. (Meg) Craig

The 1990 James A. Waight Award was given to Dr. Victor Gonzalez, a member of the BAS Board from 1971 through 1973. In 1981 he was again elected, becoming President in 1986. He held that position until 1990 when he resigned to become Chief Environmental Officer in the Ministry of Tourism and the Environment.

The 1990 James A. Waight Award was also given to Alice M. (Meg) Craig, who served on the BAS Board as Vice President from 1970 to 1987 and was a member of the Cockscomb and Education Committees. She participated in all the Society's field trips, acting as recorder, and also took part in the Christmas Bird Counts. Mrs. Craig is particularly interested in youth education.

At the same awards ceremony, Flor David received a gift in recognition of her many years service as BAS Treasurer.

1992—Janet Gibson and Gregory Smith

Janet Gibson, former BAS President, was given the James A. Waight Award for her work as Chairperson of the Reef Preservation/Half Moon Caye Committee and Coordinator of the Coastal Zone Management Project.

Gregory Smith received the James A. Waight Award for his work in marine turtle conservation. He has organized volunteer patrols of turtle nesting beaches on Ambergris Caye and at Manatee Bar.

Ford Young

Meg Craig (*center*) with Dora Weyer (*right*) and a shop owner at Mile 31, Western Highway, in front of Ford Young's Land Rover that was used for many field trips.

BAS Archives

Gregory Smith (*left*) on a turtle nesting patrol. BAS Executive Director Dolores Godfrey is second from the right.

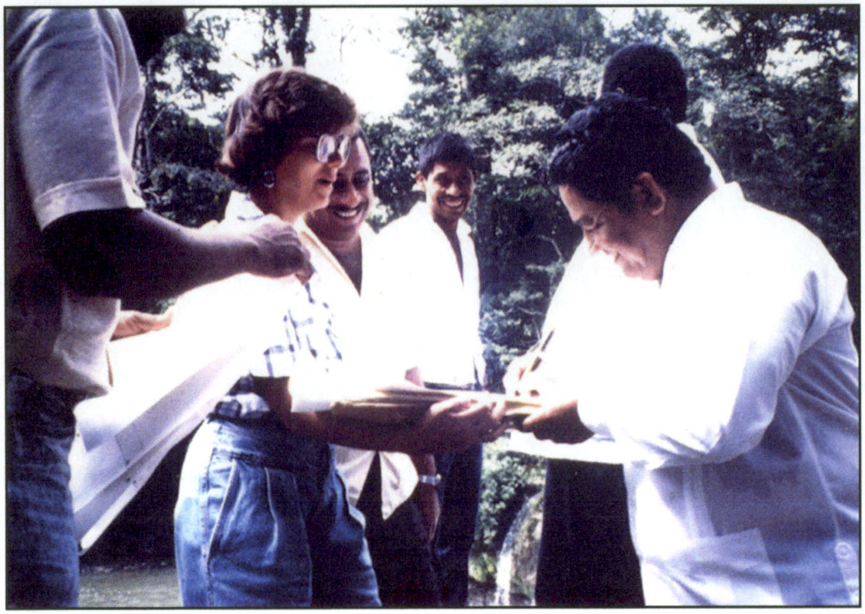

BAS Archives

Hon. Florencio Marin signing the S.I. that established the Guanacaste National Park in 1990. Therese Rath holds the papers.

1994—Hon. Florencio Marin, H.E. Dean Lindo, Rudoph Castillo, James Hyde, and Robert Leslie

The 1994 James A. Waight Award was given to several Government of Belize officials who have contributed greatly to the conservation of Belize's natural resources.

The conservation laws (Wildlife Protection and National Parks System Acts) were enacted when Hon. Florencio Marin was Minister of Natural Resources and he was responsible for establishing many reserves. In 1990 he extended the Cockscomb Basin Wildlife Sanctuary to its present acreage.

His Excellency Dean Lindo, in his term of office, established Hol Chan, Belize's first marine reserve, and several others.

Rudolph (Rudi) Castillo held several important posts as a civil servant. At different times he was Permanent Secretary in the Ministries of Education and Natural Resources, Secretary to the Cabinet and Chief Information Officer. His latest assignment was Belize's High Commissioner to Canada.

James Hyde was Commissioner of Lands and Surveys, Permanent Secretary in the Ministry of Natural Resources and later was Belize's Ambassador to the United States.

Robert (Bobby) Leslie served as Permanent Secretary in the Ministry of Foreign Affairs and was Secretary to the Cabinet. He was at one time Ambassador to the United Nations and more recently held the post of Belize's High Commissioner to London.

These three civil servants/diplomats, as President Therese Rath put it, "worked behind the scenes" during the early years of the Society and helped BAS in many important ways. It was fitting that they be honoured along with the two Ministers.

1995—G. Winston Miller

The 1995 James A. Waight Award went to G. Winston Miller, a founding member of the BAS. As Fisheries Administrator he directed manatee and sea turtle protection projects. Mr. Miller assisted greatly with the development and management of the Half Moon Caye Natural Monument by providing transportation for the Society's various projects on Half Moon Caye.

1996—Lou Nicolait

Lou Nicolait was given the James A. Waight Award in recognition of her outstanding contributions to the cause of conservation in Belize. She was elected to the BAS Board in 1984 and was an active member through 1987. She founded the Belize Center for Environmental Studies and worked closely with BAS. The *Belize Country Environmental Profile* was prepared by "Miss Lou" and her associates.

1998—Ernesto Saqui

The former Park Director of the Cockscomb Basin Wildlife Sanctuary, Ernesto Saqui, was presented the James A. Waight Award. This year BAS began having this Award Ceremony on James A. Waight's birthday, February 16. The Award was presented to Ernesto Saqui by Lydia Waight, at the Belize Marine Terminal and Museum in Belize City.

The Award was granted to Ernesto Saqui, an outstanding leader in his own right, particularly because of his invaluable contributions in changing the attitudes of the villagers of Maya Center who, in the beginning, saw conservation as a foreign idea. The award also recognized his outstanding and steadfast contributions to develop the Cockscomb Basin Wildlife Sanctuary into one of Belize's most popular and internationally renowned protected areas.

Courtesy Friends of Swallow Caye

West Indian Manatee (*Trichechus manatus*)

CBWS Park Director and James A. Waight Awardee Ernesto Saqui

1999—Julian Cho

The 1999 James A. Waight Conservation Award was presented posthumously to Julian Cho by David Craig on behalf of the Waight family. A profile of the honoree was given by Lisel Alamilla and a video presentation prepared by Great Belize Productions showed news clips of Julian Cho. The award was graciously accepted by his widow, Maggie Cho.

At the same ceremony a Junior Achievement Certificate was given by BAS to fifteen-year-old Edwin Francis Colon of Punta Gorda. Valerie Woods, Director of the Belize Tourist Board made the presentation.

2000—Sibun Watershed Association

Sibun Watershed Association was given the James A. Waight Award for their work in protecting the Sibun River Watershed and for their educational outreach programs, not to an individual, but a grassroots organization. In 1994 heavy flooding and sedimentation caused by careless land clearing practices led concerned citizens to form an organization to protect the ecosystem. The Sibun Watershed Association was formally registered as an NGO in 1997.

2001—Henry Fairweather

Henry Fairweather, BAS Founding Member, was given the James A. Waight Award for his untiring effort and environmentally friendly methodology in replanting mahogany trees and his work with the Belize River Valley Development Corporation. In 1932 and 1933 Henry Fairweather worked alongside James Waight demarking the boundary between Belize and Guatemala. At the awards ceremony Mr. Fairweather paid tribute to James Waight for his work at BAS and for Belize.

Judy Lumb

(above) The 1999 James A. Waight Award for Conservation, given posthumously to Julian Cho, was accepted by his widow, Maggie Cho.

Judy Lumb

(*right*) Edwin Francis Colon (left) of Punta Gorda, winner of the first BAS Junior Achievement Award

2002—Therese Rath

Therese Rath was given the James A. Waight Award for her work in establishing the Cockscomb Basin Wildlife Sanctuary and leading BAS from a voluntary operation to a professional, sustainable organization. Armed with studies proving that the Cockscomb Basin was one of the last strongholds for the highly endangered jaguar, Therese and other distinguished local and international environmentalists successfully lobbied for the establishment of the Cockscomb Basin Wildlife Sanctuary. She then led efforts to expand the reserve not once but twice, resulting in the magnificent 127,107 acre Sanctuary we know and love today.

When Therese was elected president of BAS in 1992, BAS was primarily a volunteer operation. Therese led the development of the first BAS strategic plan, resulting in a formalized organization and the hiring several enthusiastic young Belizeans. During Therese's first term as President, international grant funds started drying up. In a classic move, she carried around a sign saying "This Park Closed Due to Lack of Funding" which she threatened to post if she did not receive assurances of support. Her strategy worked fabulously, leading to the signing of the first contractual agreement for the management of parks between the Government and BAS, the institution of entrance fees at the protected areas, the conceptualisation of the Protected Areas Conservation Trust Act, and a reinvigorated BAS.

Judy Lumb

Therese Rath speaking at the opening of new facilities at CBWS in 1996.

Anna Rossington

John William Waight Jr. presenting the award to Therese Rath in honor of his grandfather Mr. James A. Waight.

2003—Lionel (Chocolate) Heredia

Lionel (Chocolate) Heredia was given the James A. Waight Award for his work in the protection of manatees leading to the establishment of Swallow Caye Wildlife Sanctuary. In 1978 it was reported to James Waight, the Society's founding president, that Lionel Heredia conducted regular visits to Bird Caye, a bird sanctuary managed by the Audubon. Chocolate expressed his concerns for the migrant birds and the illegal hunting activities around the Caye. After some deliberations with the Forestry Department, BAS President James A. Waight issued a statement, appointing Lionel Heredia as an Honorary Warden on Bird Caye, Northern Lagoon.

When tourism came to Caye Caulker, Chocolate became the first water taxi captain, launching a whole new industry. It was while guiding researchers that Chocolate discovered that manatees inhabited the Swallow Caye area. For many years he kept that secret to himself, but eventually other tour guides began to take tourists to see the manatees.

Chocolate was concerned for the manatees and launched an extensive advocacy campaign to protect the manatees from the increasing numbers of tourists visiting the Swallow Caye site. His persistence paid off, as Swallow Caye Wildlife Sanctuary was officially designated on July 10th, 2002.

The award was presented to Chocolate by ten-year-old Gerard Waight, grandson of James. A. Waight, who told Chocolate, "I want to be like you when I grow up."

2004—Luis Garcia

Luis Garcia was given the James A. Waight Award for his work developing an environmentally oriented radio station, which led to the formation of Belize Community Service Alliance and the international work of cleaning up the Macal, Mopan and Belize Rivers. At his young age of 30 years, Luis Garcia has already made tremendous contributions to the preservation of Belize's natural resources, community development and international relations. Using his Radio Ritmo radio station in San Ignacio Town, he started the Save our Rivers Campaign, which creates awareness, educates, sparks and assists communities with programs to conserve our natural resources. Annually, volunteers, staff, and community participants clean the banks of our rivers, the forest trails, roadsides, and other public areas, and restore riverbanks and plant trees. The Macal, Mopan and Belize Rivers transcend international borders, so these clean-up programmes were launched in cooperation with Guatemala. The two campaigns grew so rapidly that Luis formed the Belize Community Service Alliance (BCSA) and now devotes 100 percent of his time, effort and energies to BCSA.

At the same ceremony, Valdemar Andrade was awarded a long service award. CBWS Ernesto Saqui read a tribute to Mr. Andrade and Protected Areas Manager Nellie Catzim presented him a framed photograph of Half Moon Caye in appreciation of more than ten years of enthusiastic dedication to BAS.

(*above*) Maria Teresa Waight, grand-daughter of James A. Waight presents the award to Luis Garcia. (*below*) Nellie Catzim presents an award to Valdemar Andrade for his long service to the Society.

2005—Eloy Cuevas

The 2005 James A. Waight Award was given to Eloy Cuevas, a commercial fisherman, for his work in developing alternative livelihoods not only for himself but for other fishers in the southern community.

As a hunter and a fisher he recognized degradation of Belize's natural resources. He was particularly concerned about unsustainable fishing practices like the use of gillnets in southern Belize, which was quickly destroying the finfish populations. This led him to the lesser impact catch and release fly fishing, so he became one of the first fly-fishing guides in southern Belize in the early 1990s.

It was Eloy Cuevas' knowledge of Gladden Spit as an important spawning ground for the Cubera snapper that led to the declarations of the Gladden Spit and the Port Honduras Marine Reserves. Today, he advocates for the inclusion of fishers in the management of marine protected areas and has served on the Boards of the Toledo Institute for Development and Environment (TIDE) and Friends of Nature.

photos of Eloy Cuevas courtesy of Toledo Institute for Development and the Environment

2006—Pulcheria Teul

For Pulcheria Teul, the 2006 recipient of the James A. Waight Conservation Award, building human capacity in the poorest district in Belize is not a programme or project, but a life-long commitment to development of the people of Toledo and the country of Belize.

In a culture that is primarily male dominated, Pulcheria has risen from humble beginnings to become one of the leading and well recognized advocates for sustainable human development and gender equity. In 1997 when the proposed paving of the Southern Highway raised concerns about the impact upon the local communities, Pulcheria Teul was a founding member of the Environmental and Social Technical Assistance Project (ESTAP) and Programme Officer of the Southern Alliance for Grassroots Empowerment (SAGE). Through the advocacy efforts of SAGE, communities of the Toledo District achieved full participation in ESTAP with full voting rights for women representatives.

SAGE then addressed other issues impacting the Toledo District, in particular, the sustainable use and management of forest resources. SAGE engaged the communities through consultations and developed education programmes on the issue of timber extraction. They successfully lobbied the Forest Department until a Memorandum of Understanding was signed between the two.

Pulcheria was also one of the founding members of the Toledo Maya Women's Council (TMWC), formed in 1997 and became the Programme Director in 2003. She spearheads the organization of women's groups, fund-raising for capacity building, handicraft enhancement, and food and nutrition programs. Seven communities are benefiting from the work of the TMWC.

(*above*) The award was presented to Pulcheria Teul by Christopher Waight, grandson of James A. Waight.
(*below*) Pulcheria Teul with members of the Waight family

Pulcheria was the Field Coordinator for PACT's Community Co-management Project under which she mobilized 11 buffer communities linked to Aguacaliente Wildlife Sanctuary (AWS). Today, Pulcheria continues training community groups in leadership, management, planning, advocacy, and health issues, all to build the capacity of women.

2007—Sir George Brown

BAS awarded Sir. George Brown, former Chief Justice of the Supreme Court, the James A. Waight Award for his contributions to the social, cultural and environmental development of Belize, particularly through his work with youth. Sir George Brown is well known for his appointment as Chief Justice of the Supreme Court where he served until retirement in 1999.

Sir George is also known for his love for sailing, and is reported to be the leading figure behind the 9th of March Regatta. Behind the lofty title and sailing mast, lies an outstanding individual who has worked tirelessly at the grassroots level, positively influencing the lives of youth through teaching the art of sailing and sharing his love for the sea. As well, Sir George has played a critical role in protecting the constitutional rights of Belizean as he made a landmark decision that ensured that villagers of San Jose Palmar retained the rights to their land.

Today Sir George continues his work through the Sir George N. Brown Foundation and under his "Youths at Risk Programme" by training youths in athletics and coordinates technical skills projects such as boat building and sailing, and football that helps in building self-discipline, coping skills for life and preserves the Creole culture of Belize.

2008—Philip Balderamos

The Belize Audubon Society presented the 2008 James A. Waight Conservation Award to Phillip Balderamos of Belmopan City. Mr. Balderamos was awarded for his commitment to bird conservation which spans some 25 years, his unwavering advocacy and educational efforts for sustainable natural resources management, and for his contribution to community involvement in conservation efforts.

(*top*) Philip Balderamos receives the James A. Waight Conservation Award from Daniel Waight, grandson of James Waight.

(bottom) Attending the ceremony were (from right) Mrs. Balderamos, mother of the awardee, Hon. Dolores Balderamos Garcia, sister of the awardee, and Gloria and John Waight, cousin of the awardee. BAS Treasurer Mark Hulse and BAS President Earl Green are seen in the background.

Philip Balderamos enjoying the ceremony with his wife (*top*) and enjoying his new BAS cap, a gift from BAS at the award ceremony (*bottom*).

In receiving his award, Mr. Balderamos recalled his early involvement in conservation efforts as a young man and the life lessons that he was taught by the pioneers of local conservation. He encouraged Belizeans to get involved in outdoor activities to further appreciate Belize's natural resources and urged all, particularly youths, to contribute to the conservation of these resources. In coordination with BAS he was instrumental in organizing the first binocular and bird field guide distribution initiative to 100 Belizeans from all over the country.

Phillip Balderamos joined the BAS in 1981 and became involved in what is now his lifelong hobby, birding. He participated in many BAS birding field trips along with, and learned from pioneers such as, Ford Young, Meg Craig, and Dora Weyer. Mr. Balderamos served on the Board of Directors of BAS for several years and, as President in 1990-1, helped guide the transition of the Society from a purely voluntary organization to an effective and professionally managed conservation organization dedicated to maintaining a balance between people and the environment.

Since 1995, Philip has worked with the United Nations Development Programme (UNDP) as Coordinator of the Global Environment Facility Small Grants Programme (GEF SGP) in Belize. Through this work he has helped support many NGOs and community groups in conservation efforts, sustainable resource use, improved livelihoods and community empowerment.

He helped to mobilize Mayan community leaders in Toledo to advocate for the establishment of the Golden Stream Corridor Preserve (GSCP), a 15,000-acre private protected area along the Golden Stream watershed. He served as the founding president of GSCP, now Ya'axche' Conservation Trust.

2009—Janelle Chanona

Journalist Janelle Chanona received the James A. Waight Conservation Award for her outstanding environmental reporting which has raised awareness and environmental consciousness in Belize. A profile of her work at Channel 5 was given by the former owner of the station, Stewart Krohn, after which a series of video clips demonstrated the high quality of Janelle Chanona's reporting.

Several special presentations enlivened the ceremony. The life of James A. Waight was profiled in a creative dramatic presentation by Leroy Green and Leela Vernon sang and danced with the Jabiru mascot. Jackie Castillo joined Mark Phillips jazz trio to entertain with her beautiful voice.

(*top*) Ellen Waight DelMargo, daughter of James and Lydia Waight, presents the James A. Waight Conversation Award to Janelle Chanona.
(*above*) BAS President Earl Green congratulates Janelle Chanona.
(*left*) Leela Vernon singing and dancing with Jimbo Roo, the Jabiru mascot.

(*above from left*) Gloria Waight, Christopher Waight, Daniel Waight, Ellen Waight DelMargo, Sir Dr. Colville Young, Govenor General of Belize, Janelle Chanona, John Waight, Gerard Waight, BAS Executive Director Anna D. Hoare.

(*below*) James A. Waight was profiled in an excellent dramatic presentation by Leroy Green. (*photos by Judy Lumb*))

Ellen Waight DelMargo presents the award to Dr. Ed Boles.

2010—Dr. Ed Boles

Raymon Edward (Ed) Boles, Ph.D., is the 2010 recipient of the James A. Waight Conservation Award. Originally from Mississippi, Dr. Boles has been in Belize for more than 20 years. His doctoral dissertation was entitled "Development of a Community-Based Rapid Watershed Assessment Protocol focusing on the Sibun River Watershed, Belize." He was instrumental in the formation of the Sibun River Watershed Association, which was the winner of the 2000 James A. Waight Conservation Award. Among his many publications is the impressive *Sibun River Watershed Atlas.*

Dr. Boles has devoted his life to freshwater aquatic biology and ecology, and has inspired many students while teaching at the University of Belize (UB) and Galen University. At UB he spearheaded the development of the Bachelor's programme in Natural Resource Management, obtaining funding from PACT and the Oak Foundation for programme and facilities development. His fieldwork includes conducting environmental assessments for development projects, as well as biodiversity, conservation, and management assessments throughout Belize.

If all this scientific work were not enough, Dr. Boles is also an artist and writer. He has written children's stories to create environmental awareness as well as entertain. For all these and many more accomplishments in a life devoted to conservation, the 2010 James A. Waight Conservation Award was presented to Dr.

(*from left*) Ellen Waight DelMargo, Richard DelMargo, Madeleine Guerra, Yael Perera (modeling the black garbage bag dress), Francesca Sharp (with green bags), Dr. Ed Boles, Jasmine Skeen (in bubble-wrap dress), and Nikisha Ack (rainforest dress).

Ed Boles by Ellen Waight DelMargo, daughter of James A. Waight.

A special presentation of dance and poetry by Madeleine Guerra entitled "350 Steps to Climate Change" and an Eco-Fashion Show enhanced the ceremony. Madeline Guerra has been a very active young BAS member as she organized a large group to participate in the Earth Walk and contributed one thousand dollars towards BAS's Education Fund that is used to assist needy buffer community students.

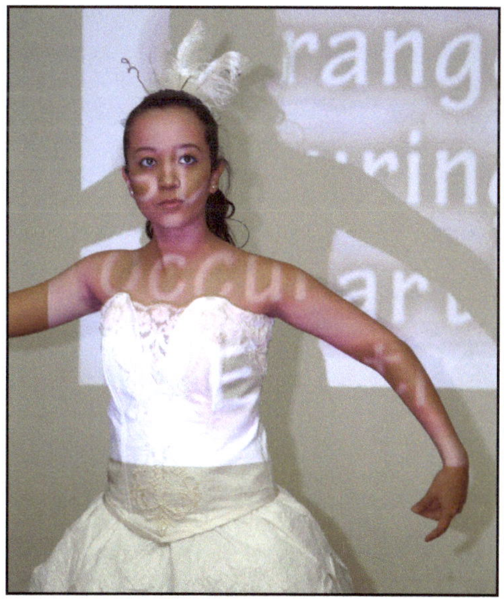

Madeleine Guerra in "350 Steps to Climate Change"

World Wildlife Fund Award

During his visit to the Cockscomb Basin Wildlife Sanctuary on February 6, 1988, H.R.H. Prince Philip, Duke of Edinburgh, presented Ignacio Pop, Senior Warden at the Cockscomb Basin Wildlife Sanctuary with the World Wildlife Fund Award for his work at the Sanctuary. Mr. Pop was first a Forest Ranger and then became a Warden when the Sanctuary was declared.

Warden of the Year

Ignacio Pop

Ignacio Pop and Steve Tillett were both recipients of BAS' first Warden of the Year Awards, presented to them at BAS' 29th Annual General Meeting (AGM) on May 26, 1998. The awards were given in recognition of their outstanding performance during their 14 years of service with the Society.

Ignacio Pop (front) with HRH Prince Philip (right) in 1988.

Ignacio Pop was one of the first wardens hired to work at the CBWS, even before the Society became officially involved in its management in 1986. He has worked continuously at the Sanctuary since that time. In 2005 Ignacio Pop was given a Loyalty Award for 20 years of service. After 26 years working in the CBWS, Mr. Pop retired in 2009. He was honoured with the Long Service Award at the 2009 James A. Waight Award Ceremony.

Steve Tillett

Ever since he was hired in the formative years of the CTWS, Steve Tillett has performed with dedication. He has participated in numerous specialized wildlife conservation training sessions and is one of the most lively and knowledgeable staff naturalists. He was the recipient of the first Warden of the Year Award in 1998, along with Ignacio Pop. He was given the Loyalty Award in 2005 after 20 years as a Warden in the Crooked Tree Wildlife Sanctuary.

BAS Protected Areas Manager Dominique Lizama presents the Long Service Award to Ignacio Pop.

Eugenio Ah

The second Warden of the Year Award was presented at the 30th BAS AGM to Eugenio Ah for his outstanding and dedicated service to the Society.

Patrick Scott, Jr.

At the 31st AGM Patrick Scott, Jr., was given the third Warden of the Year Award and Pamela Scott was given a certification of appreciation by all the park directors for her work as Protected Areas Manager.

Emiliano Sho

At the 32nd AGM on April 30, 2001, Emiliano Sho was given the Warden of the Year Award.

Israel Manzanero, Sr

Israel Manzanero, Sr., was given the Warden of the Year Award at the 33rd AGM in 2002. The award was presented by Sergio Hoare, Marine Protected Areas Coordinator.

Nicacio Coc

At the 34th AGM. Nicacio Coc was given the Warden of the Year Award, saying that he is reliable, hard-working, and truly believes in the mission of BAS.

Volunteer of the Year

Dirk Francisco

In 2001 Dirk Francisco was awarded the only Volunteer of the Year Award that BAS has given in the first 40 years for his outstanding work in Environmental Education and in general support of the work of the Society.

Emiliano Sho with his Warden of the Year Award in 2001

Sergio Hoare presents Warden of the Year to Israel Manzanero, Sr.

Steve Tillett receives the Loyalty Award in 2005 after 20 years as Crooked Tree Wildlife Sanctuary Warden

Employee of the Year Award is presented to Heron Moreno by Executive Director Valdemar Andrade.

Other BAS Awards

At the 34th AGM Heron Moreno was given the first BAS Employee of the Year for his steadfast and enthusiastic work as the BAS Community Liaison Officer. First Vice President Earl Green was given the first BAS Board Member of the Year Award also at the 34th AGM in 2003.

At the 37th AGM outgoing President David Craig and First Secretary Judy Lumb were given plaques in recognition of their service to the Society.

At the 38th AGM awards were presented to the members of the Board of Trustees: David Craig, Barry Bowen, Mary Pearl, and Tony Flynn. Partnership Awards were given to the Forest Department, which included a Gold Membership to the Society, and to the Belize National Coast Guard Service, which included a Platinum Membership.

Outgoing First Secretary Nicole Auil was honoured at the 39th AGM for her long service to the Society.

Mrs. Lydia Waight (*center*) points out birds to visitors on a 1970 field trip to Crooked Tree Lagoon.

Lydia Waight, Knight of the Order of the Golden Ark

Belize National Awards

On December 12, 1990, the Honourable Florencio Marin, Minister of Natural Resources, presented awards to founding members W. Ford Young, Dora Weyer, Alice M. Craig, James A. Waight and Lydia Waight. Mr. Marin commended BAS for "lobbying over the years for the conservation of Belize's natural heritage."

Goldman Environmental Prize

The Goldman Environment Prize was established in 1990 by an international committee. It has the prestige of a Nobel Prize and is awarded to "men and women of vision and courage, who take great risks for the environment." Former BAS President, Janet Gibson was one of the six "grassroots heroes" from around the world who were honoured with the first Goldman Environment Prizes in San Francisco on Earth Day, 1990. Mrs. Gibson was chosen for her work in establishing the Hol Chan Marine Reserve, the first of its kind in Central America.

Knight of the Golden Ark

In November of 1996, His Royal Highness Prince Bernhard of the Netherlands honoured BAS founding member and secretary Mrs. Lydia Waight with the award of the "Knight of the Order of the Golden Ark." Mrs. Waight was granted the award because of her years of seemingly tireless effort towards conservation. For 27 years she was a driving force behind the Belize Audubon Society. Her example of sincere volunteer service serves is a guiding example to the entire conservation community in Belize and throughout the world.

During National Tourism Week of 2004, the Belize Tourism Board awarded BAS the Environmental Organism of the Year Award recognizing excellence in our work.

In 2008, the Family Life Association gave BAS an award in recognition of our invaluable contributions to the success of their youth activities.

The Maya Center community provided BAS with a certificate based on our generous contribution towards the construction of the monument and assistance to the village in November 2009.

Sacred Heart College provided a certificate of appreciation for BAS' contribution to their work in December 2009.

In 2009, BAS was awarded a certificate of appreciation given by the Deputy Prime Minister and Minister of Natural Resources Hon Gaspar Vega and the Forest Department for our 40 years of outstanding contribution to Protected Areas Management in Belize.

In 2009 BAS received an award based on the outstanding support to the Coast Guard from Commandant John Borland from the National Coast Guard.

SUMMARY OF AWARDS

The James A. Waight Award Ceremony is held each year on February 16, the birthday of James A. Waight, BAS Founder and President for the first 16 years. The Ceremony is always a wonderful occasion, one of the highlights of the year for BAS members. The Award has been given by BAS to the following conservationists:

1987 Richard Charles Atkinson
1988 W. Ford Young and the Community Baboon Sanctuary
1989 Dora Weyer
1990 Dr. Victor Gonzalez and Alice M. (Meg) Craig
1992 Janet Gibson and Gregory Smith
1994 Hon. Florencio Marin, H.E. Dean Lindo, Rudoph Castillo, James Hyde, and Robert Leslie
1996 G. Winston Miller
1996 Lou Nicolait
1998 Ernesto Saqui
1999 Julian Cho
2000 Sibun Watershed Association
2001 Terese Rath
2002 Henry Fairweather
2003 Lionel (Chocolate) Heredia
2004 Luis Garcia
2005 Eloy Cuevas
2006 Pulcheria Teul
2007 Sir George Brown
2008 Philip Balderamos
2009 Janelle Chanona
2010 Dr. Ed Boles

BAS as an organization and BAS members have also received awards. Former BAS President Janet Gibson was awarded the prestigious international Goldman Environment Prize in 1990. In 1996 Lydia Waight, BAS Founder and Secretary for the first 27 years, was awarded the Knight of the Order of the Golden Ark by His Royal Highness Prince Bernhard of the Netherlands

BELIZE AUDUBON SOCIETY

Celebrating 40 Years

Chapter Seven

In Appreciation of Our Partners

We hope that all our members and friends are constantly aware of our deep appreciation for the support we have received through the present time. It is appropriate at the time of our 40th Anniversary to say thank you to all who have helped make the Society what it is today. The efforts of local members and organizations have made major contributions to BAS' success. We have been able to triple the impact of these efforts using the technical and financial assistance of our international partners. While it is our intention to give recognition to all those who have contributed, we apologize to anyone we might have inadvertently omitted.

For 15 years the Belize Audubon Society was the only environmental organization in Belize. By the end of the 1980s, BAS had the pleasure of endorsing the formation of the Belize Center for Environmental Studies, the Belize Zoo and Tropical Education Center, and Programme for Belize, which formed the basis of Belize's cohesive environmental community.

Belize Center for Environmental Studies—Toledo Institute for Development and Environment

In the fall of 1984 the Belize Environmental Country Profile was released. It had been prepared by Robert Nicolait and Associates under a contract from the United States Agency for International Development. This contract formed the basis for the formation of the Belize Center for Environmental Studies (BCES) under the direction of Lou Nicolait. For the next twelve years the BCES worked closely with BAS and other organizations in Belize on a variety of environmental studies. Most notable of these studies was their focus in the Toledo District on a large area of contiguous protected areas from the ridge of the Maya Mountains to the sea. When the BCES closed its doors in 1997, responsibility for this important work was taken over by Wil Maheia who founded the Toledo Institute for Development and Environment (TIDE).

Belize Zoo

The Belize Zoo, developed by Sharon Matola, has always been committed to conservation. The Belize Zoo developed into the Belize Zoo and Tropical Education Center which makes many contributions to conservation in Belize, especially in the training of tour guides. They have collaborated with BAS on several projects, including the highly successful Toledo and Corozal Environmental Carnivals, which also involved Programme for Belize.

Programme for Belize

Programme for Belize was officially announced by the Massachusetts Audubon Society in December of 1987 to make an impact upon the development of Belize. Programme for Belize manages the Rio Bravo Conservation and Management Area and other environmental conservation projects. Programme for Belize has cooperated with BAS on a number of environmental projects, especially during environmental crises such as the appearance of tilapia in northern Belize and the killing of scarlet macaws in Red Bank Village.

Green Reef-Belize

Green Reef-Belize a private, non-profit membership organization dedicated to the promotion of sustainable use and conservation of Belize's marine and coastal resources was founded in 1996 by a group of concerned citizens and residents of San Pedro, Ambergris Caye. In 1998 BAS signed a Memorandum of Understanding with Green Reef-Belize for the management of two bird sanctuaries located near Ambergris Caye. BAS facilitated the production of Green Reef's Strategic Plan, assisted with development of funding proposals, and co-sponsored the application for a volunteer from the Peace Corps.

Community Baboon Sanctuary

The Community Baboon Sanctuary (CBS) was established in 1985 to protect one of the few healthy black howler monkey (*Alouatta pigra*) populations in Central America. The Sanctuary is a completely voluntary grassroots conservation programme. In 1987 grants were received from the World Wildlife Fund and the Zoological Society of Milwaukee County. In 1990 a grant from Inter-American Foundation was awarded to establish a revolving fund from which loans were made to persons in the Sanctuary desiring to develop small tourist accommodations, all of which BAS administered. In 1996 BAS ended its period of oversight and the CBS has operated independently ever since.

Shipstern Nature Reserve

Shipstern Wildlife Reserve and Butterfly Farm was established in 1987. The area of some 22,000 acres in the Corozal District includes northern hardwood forest, saline lagoon systems and mangrove shoreline. In 1994 the International Tropical Conservation Foundation (ICTF) of Neuchatel, Switzerland, and the Belize Audubon Society signed a contract for the management of the privately-owned Shipstern Nature Reserve. Under the contract, BAS managed the reserve with the ICTF providing the funds. This arrangement continued until 1997.

BAS Archives

Donald Tillett (left) accepts management of the Shipstern Wilidlife Sanctuary on behalf of BAS from Jan Meerman

Belize Alliance of Conservation Non-government Organizations

The need for an official alliance of conservation NGOs was first felt in 1993 when the Natural Resources Management Project (NARMAP) asked for one representative from the environmental NGOs to serve on their Technical Advisory Committee. The Belize Alliance of Conservation Non-government Organizations (BACONGO) was formed to select that representative. Soon after its formation BACONGO became a united lobbying group when a proposal for logging in the Columbia Forest Reserve was presented to Hon. Florencio Marin, Minister of Natural Resources by a Mexican firm. BACONGO expressed concerns about this proposal and continued to serve as a vehicle for the environmental community to present a united front in the face of environmental threats.

In 1999 BACONGO membership included BAS, Programme for Belize, the Belize Zoo and Tropical Education Center, Sibun Watershed Association, Toledo Institute for Development and Environment, Association of Friends of Freshwater Creek, Friends of Gra-Gra Lagoon, and Green Reef. BAS had difficulty functioning as a part of BACONGO because BAS is a membership organization that is headed by an elected Board of Directors. Positions on developmental and other issues have to be approved by the BAS Board or Directors. The representatives from BAS could not endorse positions on the spot at BACONGO meetings until the BAS Board had a chance to study the situation. After numerous unsuccessful attempts to change the BACONGO decision-making process, BAS regrettably withdrew its membership.

Association of National Development Agencies

The Association of National Development Agencies (ANDA) was established as an umbrella organization to support grassroots organizations that are working for socioeconomic change and self reliance; to improve interaction between Belizean developmental agencies, national and international NGOs; and to encourage development that promotes and preserves a Belizean identity. BAS joined ANDA in late 1997 and held the Secretariat for several years. BAS represented ANDA on the National Environmental Assessment Committee (NEAC) from the beginning of the committee until 2007 when BAS was appointed as a member of NEAC.

Association of Protected Areas Management Organizations

Recognizing the problems and challenges resulting from inadequate and ineffective coordination among the protected areas management organizations, BAS and UNDP-Belize spearheaded a meeting on 23 January 2003 in which the Association of Protected Areas Management Organizations (APAMO) was created. APAMO has been advocating for maintaining the integrity of the protected areas and for the PACT amendments to ensure better financing of protected areas. APAMO developed a proposed co-management framework and has been lobbying with appropriate Ministers of the Government of Belize for its adoption and passage of appropriate legislation. In 2010 APAMO began a public awareness campaign against drilling for oil offshore and in protected areas.

Mario Teul (*holding bird*) and Alberto Yanosky of Birds without Borders instruct students in 2005 for the Bird Fest at Crooked Tree Wildlife Sanctuary.

Other Belizean Organizations

Belize Enterprise for Sustainable Technology

BAS collaborated with Belize Enterprise for Sustainable Technology (BEST) for an empowerment training programme in the buffer communities surrounding BAS-managed protected areas to develop capacities. In 2001 BEST and BAS protected areas personnel conducted two programmes, "Leadership for Community Development", and "Sustainable Livelihoods", and helped with business plans for buffer community projects.

Belize Fishermen's Cooperatives

A Memorandum of Understanding was signed on April 5, 2004, between the Belize Audubon Society (BAS) and the Belize Fishermen's Cooperatives Association (BFCA) to work in partnership for the sustainable use of the resources of Lighthouse Reef Atoll to ensure that the full benefits to all stakeholders are realized.

One outcome of this partnership was the research and lobbying effort that resulted in the protection of the spawning aggregation sites for Nassau groupers (*Chapter Five*).

Birds without Borders

Since 1998 Birds Without Borders–Aves Sin Fronteas (BWB-ASF) has been doing field research in Belize coordinated by the Zoological Society of Milwaukee (Wisconsin, USA) and the Foundation for Wildlife Conservation, in cooperation with private landowners and local schools. BWB-ASF research results are used to help conserve Belize's resident and migratory birds by providing recommendations to landowners on how they can manage their land to benefit birds.

Marine Welding

Ian Courtney and the staff of Marine Welding donated an upgrade of the ladder for the observation deck at the booby rookery at Half Moon Caye.

Pine Lumber

Amin Bedran of Pine Lumber arranged for a new BAS Gift Shop to be built at the Fort Street Office in Belize City to take advantage of cruise ship visitors that frequent the area. The cost of the building was repaid out of the proceeds from sales.

Rare Animal Relief Effort

The Rare Animal Relief Effort (RARE) has supported BAS efforts for a number of years. In 1993 BAS and RARE collaborated on an educational campaign that focussed on Rainbow, the toucan mascot.

In 2000 RARE provided support for the development of a interpretive trail, the RARE Trail in the Blue Hole National Park, later renamed St. Herman's Blue Hole National Park. From 2007 to 2009 RARE supported a major educational campaign with the buffer communities of the Crooked Tree Wildlife Sanctuary.

Belize National Coast Guard

This government agency was established in 2005 with responsibilities to enforce all maritime laws, protect resources, ensure safety at sea, conduct operations at sea, and support the departments with mandates to enforce laws at sea. They have always assisted BAS in our work.

(*above*) BAS Gift Shop built in 2003 by generosity of Pine Lumber

(*below*) Rainbow mascot in 1993 RARE-BAS educational campaign

BAS Archives

International Organizations

BirdLife International

In 1980, Belize was invited to become a member of the International Council for the Bird Preservation. The membership was to be through BAS because Ford Young was to be the representative. BAS was accepted as a member, but unable to pay the membership fees. This Council was later changed to BirdLife International, and in 1998, BAS became a Partner Designate.

BirdLife International is a global alliance of conservation organizations with a focus on birds. Together the BirdLife partners form the leading authority on the status of birds, their habitats and the issues and problems affecting bird life. The BAS has been working with BirdLife International for many years on shared priorities, policies and programmes for conservation action; exchanging skills, achievements and information; and growing in ability, authority and influence.

From 15–20 May, 2005, BAS hosted the IVth BirdLife International Americas Partnership Meeting. The theme for day one was "Sustainable Livelihoods" and on day two attendees considered "Advancing Important Bird Areas in the Americas". On day three field trips were offered to St. Herman's Blue Hole National Park and Half Moon Caye and Blue Hole Natural Monument. During the field trip to Half Moon Caye, a new bird for Belize was seen, the White-throated Sparrow. On day four the theme was "Conserving and Monitoring Threatened Species".

BirdLife International Americas Partnership Meeting Opening Ceremony 16th May 2005 at House of Culture:
(*top*) Keynote Speaker Hon. John Briceño presents an award to Ben Olewin for his work with Important Bird Areas.
(*bottom*) Alberto Yanosky, Chair, BirdLife Americas Regional Council, welcomes everyone to the meeting.

BAS Hosts BirdLife International Americas Partnership Meeting 16th-20th May, 2005 Belize City, Belize

Attendees at the Opening Ceremony of the BirdLife International Americas Partnership Meeting (*below*). The Belize Dance Company (*above* and *right*) and Pablo Collado (*left*) provided entertainment.

Themes for each day of the BirdLife Meeting included:

- Sustainable Livelihoods,
- Advancing Important Birds Areas, and
- Conserving and Monitoring Threatened Species.

One of the highlights of the Meeting was during the field trip to Half Moon Caye when a new bird species for Belize was identified, the White-throated Sparrow (*Zonotrichia alicollis – below, right*).

Kerem Boyla

BirdLife International supported a project at Crooked Tree Wildlife Sanctuary entitled "Garnering support for Crooked Tree Wildlife Sanctuary by creating opportunities through capacity building for the local community to benefit from conservation practices".

Eden Conservancy / Target Earth / Jaguar Creek

After a lengthy negotiation, the Eden Conservancy, through their Target Earth programme, donated 3,300 acres of land to the Government of Belize to be added to the SHBHNP, which connects it with the Sibun Forest Reserve. Jaguar Creek, also owned by Eden Conservancy, has contributed to the Armenia Village Women's Group.

Happy Hollow Zoo / Jaguar Conservation Trust

In 2001 the Happy Hollow Zoo established a donation programme with the funds going to the Jaguar Conservation Trust and informed BAS that the CBWS would be the recipient of these collected funds.

(*above*) Hospitality training in Crooked Tree Lagoon
(*below*) Tour in Crooked Tree Lagoon

BAS Archives

Donna Monroe

Half Moon Caye Natural Monument

Gene Wright of Old Woman Creek Nature Estuarine and Research Reserve displays educational materials donated to BAS.

Island Expeditions

Island Expeditions is a tour company that brings groups to camp at Half Moon Caye. Through the years Island Expeditions has been very supportive of the BAS management of Half Moon Caye and Blue Hole Natural Monuments.

Following Hurricane Ivan in 2003, Island Expeditions repaired the ground water plumbing for the camping area at Half Moon Caye. Through Island Expeditions support, in 2006 BAS hosted two students from the fishing communities Sarteneja and Chunox to participate in a week-long visit to Half Moon Caye Natural Monument and the Actun Tunichil Muknal cave.

Old Woman Creek

Initiated when Linda Feix visited Belize for the Coastal Zone and Ecotourism Congress in 2000, the "Birds with two Homes" initiative has grown into an outreach project between the Old Woman Creek Nature Estuarine and Research Reserve (OWC) buffer community schools and the primary schools of the CTWS buffer communities CTWS. In 2003 four teachers from the OWC buffer schools visited the schools in the villages of Lemonal, Biscayne, Maypen and Crooked Tree to see first-hand the resources available to our teachers and the structure that governs our primary education. They were impressed with the level of commitment that the teachers exhibited and donated stationery to each school.

REDES-Central America

BAS hosted the first general assembly of the Regional Network of Environmental and Non-governmental Organizations Working Together for the Sustainable Development of Central America (REDES-Centroamerica), held in Belize at the St. Catherine Academy from August 24 to 27, 1988. Each of the Central American countries was represented by ten delegates from their NGOs. Representatives of other international environmental organizations were also in attendance. An award was given to the Belize Audubon Society by REDES in recognition of its past efforts in the conservation of the natural patrimony of Belize. The closing address was made by Carlos Perdomo, Secretary of the Cabinet of the Government of Belize.

BAS Archives

(*above*) BAS officers and staff with award at REDES (back from left) Walter (Micky) Craig, Bardi Riverol, James Waight, (front from left) Ernesto Saqui, Victor Gonzalez, Lydia Waight and Fallett Young, (*below*) This award still hangs in the BAS Conference Room.

University Relationships

BAS has cultivated relationships with a number of universities and other protected areas managers, including the University of Belize, University of North Carolina at Wilmington, University of Michigan, University of Seattle, State University of New York, University of Idaho, University of Quintana Roo, Mesa Verde, and Indonesia.

World Conservation Union (IUCN)

The BAS first became a member of the World Conservation Union (IUCN) in April of 1986. The second annual meeting of Regional Meso-American Committees of the IUCN was held in Belize from October 20 to 23, 1994. BAS was responsible for planning the programme of events. Attending the meeting were national committee presidents of the Mesoamerican region from Mexico to Panama, IUCN Regional Director Enrique Lahmann and Dr. A. Cerda, lecturer in Natural Resource Economics, University of Concepcion, Chile. Among matters discussed was developing a closer relationship with the IUCN Commissions.

BAS has been the most active member of the IUCN in Belize, participating in the World Congresses in Jordan, Thailand, and Barcelona.

BAS Archives

BAS Vice President David Craig receives check from Longacre Director David Baraff in 1991

BAS Archives

Picnic table at Guanacaste National Park built by Longacre and Belizean youth volunteers in 1991

VOLUNTEERS

For the first fifteen years of BAS' existence, all the work was accomplished through the efforts of volunteers. Even after it was possible to hire professional staff, volunteers have continued to serve the Society. Many individuals have volunteered for varying lengths of time in the BAS Office and in the protected areas. In 1999, and again in 2002, the volunteer programme at BAS was organized to make it run smoothly for both the volunteer and the Society.

BAS is grateful for every volunteer and looks forward to many more as the environmental work in Belize needs many hands.

Belize Youth Conservation Corps

From 1991 through 1995 Belize Youth Conservation Corps (BYCC) worked on projects in several protected areas. In CBWS they built or repaired several bridges, improved trails, cut new trails, and dug a garbage pit. At GNP they built nesting boxes for birds, roosting boxes for bats, made hanging baskets for orchids, constructed a self-guided trail, cleaned in preparation for the opening of the Visitor Center, and repaired the steps leading down to the Belize River Recreation Center. At HMCNM they cleaned the beach and cleared trails.

Longacre Expeditions

Youths from Longacre Expeditions came to Belize in the summer of 1991. Joined by Belizean youths from Hattieville, organized by James Sanker, they built 18 picnic tables for protected areas of Belize.

LightHawk

LightHawk is a New Mexico-based conservation airwing. Since 1989 they have come to Belize each year to fly over critical areas to give passengers a view of existing pressures from population, agriculture and development on protected terrestrial and marine areas. The experience of seeing first-hand the expanse of lush forest vegetation, interrupted by its meandering rivers, is awesome and indeed educational. Even more impressive is the viewing of Victoria Peak from an arm's length distance. These passengers often include government ministers or buffer community members.

In 1994 a settlement was observed very near to the southern boundary of CBWS. Land clearing was observed within the boundary of the Bladen Branch Nature Reserve and dispersed settlements were noted not too far from the eastern boundary of the Reserve.

The LightHawk flight in 2000 showed the importance of the donation by the Eden Conservancy of over 3,000 acres that connects the Blue Hole National Park (now St. Herman's Blue Hole National Park) with the Sibun Forest Reserve.

A canopy walk was being considered for St. Herman's Blue Hole National Park in 2004, but the LightHawk fly-over showed that the area was second growth forest and not suitable for such a venture.

The fly-overs in 2006 demonstrated that the development on Black Creek by developer Luke Espat had encroached into the Crooked Tree Wildlife Sanctuary. A small rubble caye had developed east of Half Moon Caye, named "Mitch Caye" because it appeared after Hurricane Mitch.

In 2007 encroachment of the coconut trees into the littoral forest was noted on Half Moon Caye and Mitch Caye had grown considerably. A trail was seen from Ringtail Village into the SHBHNP and development was seen on three sides of TMNR.

BAS Archives

(*top*) Aerial view of Victoria Peak from LightHawk (*centre*) Crooked Tree community member Verna Samuels and BAS Executive Director Valdemar Andrade on LightHawk flight. (*bottom*) New Mitch Caye on the lower left at Half Moon Caye

Peace Corps Volunteer Scott Schmidt works with the children at the 39th BAS Annual General Meeting in 2008.

Raleigh International volunteers work on the upgrade of the MIller House on Half Moon Caye in 2000.

Peace Corps

The Peace Corps, a volunteer organization of government of the United States, has provided several volunteers, beginning with Catherine and Tom Meekin in 1985, both of whom were shared with other projects in Belize. Most recently (2006-8) Scott Schmidt spent a very productive two years supporting the Environmental Education Programme at BAS.

Raleigh International

Operation Raleigh, a British youth development charity, built a Visitor Center in Crooked Tree Wildlife Sanctuary in June of 1986. After an absence of ten years Operation Raleigh, renamed Raleigh International, returned to Belize in 1995. Raleigh venturers, including Belizeans, worked on a new trail through St. Herman's Cave, Blue Hole National Park. At Cockscomb they constructed a trail to the Outlier and built new bridges. They also worked on Half Moon Caye, building a Visitor Center and repairing the observation deck in the rookery.

Back again in 1996, Raleigh constructed the Community Resource Center at Tapir Mountain Nature Reserve. At Cockscomb they improved the Victoria Peak Trail and built a cabin with materials donated by the British High Commission.

In 2000 Raleigh volunteers worked in Half Moon Caye Natural Monument upgrading the Miller House for researchers.

Student and Scout Groups

Youth Environmental Action Group (YEAG) was formed in San Jose, Succotz. They assisted BAS with tours of the GNP during the 1989 Agricultural Fair in Belmopan. In August of 1990 BAS held a workshop in GNP for YEAG.

Sixty-five members of Students Understanding Nature (SUN) St. John's Sixth Form helped with signs and clearing trails in GNP in 1992. In 1994, 21 members of Students Encouraging Environmental Development (SEED) from Belmopan Comprehensive School conducted a cleaning campaign at GNP.

By 2000 there were environmental clubs at Corozal Junior College, Muffles Junior College, Wesley College, St. Catherine's Academy, Grace Primary School, and Belize Elementary School.

In 2009 Venture Scouts from Holy Redeemer, St. Ignatius, St. Joseph and Burrell Boom units climbed Victoria Peak. That same year BAS organized five trips to Victoria peak with people from all walks of life as part of the 40th anniversary celebration.

Swim for the Reef

In 1985 and again in 1990 Captain Ted Martin, a British soldier, swam from St. George's Caye to Belize City for the Reef Preservation Fund. In 1991 the Environmental Youth Club, assisted by British Forces Belize, continued this tradition and swam the distance in relay fashion with the five older children taking 40-minute laps and the two younger children taking 10-minute laps.

Courtesy Venture Scouts

(*above*) Venture Scouts on top of Victoria Peak

(*below*) BAS President Janet Gibson congratulates Captain Ted Martin on his "Swim for the Reef" in 1990

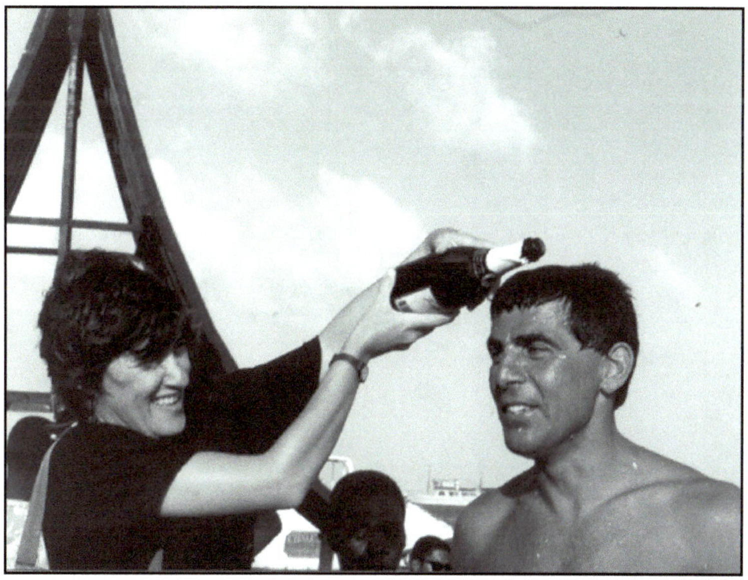

BAS Archives

Courtesy Trekforce

Trekforce volunteers improve the 12K Base Camp in 2005

Heron Moreno

World Gap Challenge volunteers Bob MacKintosh and Olly Bates working in CBWS in 2003

Trekforce

Trekforce is a United Kingdom youth development charity that has brought groups of enthusiastic British youth to Belize to work on projects in protected areas. They built the Observation Tower in Crooked Tree Wildlife Sanctuary and improved trails in the West Basin of the Cockscomb Basin in the Spring of 1998. Their Fall 1998 projects included building latrines at the Maya Mopan entrance to the Cockscomb Basin Wildlife Sanctuary, but that project was terminated just before completion because of Hurricane Mitch. In Belize City the volunteers helped with clean-up of schools that had been used as hurricane shelters while they waited for the sea to calm down enough to go to Half Moon Caye, where they also cleaned up after Mitch.

In 1999 Trekforce developed the 12-kilometer base camp on the trail to Victoria Peak. In 2003 Trekforce volunteers cleared approximately 15 km of trails and boundaries for the BAS staff to patrol Tapir Mountain Nature Reserve.

World Challenge

World Challenge is a United Kingdom-based charity that organizes work experiences for youth choosing to take a gap year before entering college. Gap Challenge volunteers have been assigned to work with the wardens at the CBWS. In 2001 Gap Challenge volunteers helped the Staff of the SHBHNP install aluminium pillars with reflective tape all along the edge of the trail 350 yards into St. Herman's Cave.

Stakeholder Communities

Since Steve Tillett became the first CTWS warden and Ignacio Pop and his son Pedro became the first CBWSwardens, BAS has hired wardens from the communities surrounding the protected areas and worked with the communities to ensure their support of the protected areas that BAS manages. The main focus of the Community development program is to offset environmental pressure on protected areas managed by BAS.

This work was greatly expanded in 2000 with the funding of the co-management project by European Union and the creation of the Community Liaison Officer (CLO) position at BAS. The first CLO at BAS was Heron Moreno from 2002–2005. Orlando Jimenez held the CLO position from 2006–2008 and Lucito Ayuso is the CLO in 2009. Local Advisory Committees were formed in buffer communities to facilitate communication and BAS work with the villages.

BAS community development work has been supported by AVINA, BirdLife International, COMPACT, HIVOS, the Jaguar Project, Oak Foundation and PACT. Partners involved in the community work with BAS include BEST, Women Development Department, Department of Agriculture, Forest Department, and the Institute of Technical Vocational Education Training.

Working in Partnership with Communities for the Conservation of Protected Areas was developed in 2008 by BAS with funds from BirdLife International. This document outlines some of the activities carried out by BAS to strengthen community relationships, build appreciation for protected areas, and facilitate alternative income generation to take the pressure off protected areas.

(*above*) In 2003 European Union Representatives visited CTWS.

(*right*) *Working in Partnership with Communities* documents BAS work in communities.

At CBWS a slate carving of the EU Logo were presented to the EU Representatives

Maya Center Women's Group in 2004

Cockscomb Basin Wildlife Sanctuary

The Maya Center Women's Group was formed soon after the Sanctuary was developed. When visitors began to come, the women in the village made crafts for sale. Soon they organized and built a gift shop at the entrance to the village. They have been very successful and have shared their experience with groups in other communties.

The co-management project worked with the CBWS stakeholder communities of Georgetown, Maya Center, Maya Mopan, Red Bank, San Pablo, San Roman, and Santa Rosa. The Maya Mopan Honey Cooperative was developed and registered by 2003. At the same time, the Local Advisory Committee of Maya Mopan reported their concerns about an illegal logging operation on the CBWS boundary and a problem jaguar that had been killing domestic livestock in the village. The Forest Department gave permission for the village to destroy the jaguar, the carcass of which was disposed of through the appropriate legal procedures by the Forest Department.

After Hurricane Iris in 2001, in cooperation with the Belize Red Cross, BAS set up a relief center in Maya Center to serve the affected areas. This gave the people of Maya Center the opportunity to be involved in the relief efforts.

BAS helped the village of San Pablo acquire coordinates, survey and secure their communal lands. The Red Bank Tourism Group asked BAS for help in organizing their tourism venture.

The Maya Mopan Cacao Farmers asked BAS for help in acquiring land and developing cacao plantations, which are considered a good buffer outside the boundary of the CBWS.

In 2005 the Maya Center community asked BAS for help addressing concerns about potential socio-economic threats from the proposed Sapadilla Lagoon Resort, Golf Course, and Casino planned for development close to the village.

A jaguar was found dead near the CBWS, which prompted BAS to hold a consultative workshop with farmers to share ideas on methologies that can be used to control jaguar predation on livestock.

Maya Center Women's Craft Shop opening of new building November 2, 2002. (*top*) Members of the Maya Center Women's Group in the audience; (*nset*) Slate carving of the BAS Logo (*bottom*) Sarita Teck shows her slate caving.

Cashew Cooperative Steering Committee

Orlando Jimenez

Community planning session with community members from La Bomba, Lemonal and Flowers Bank. BAS facilitated planning for bee-keeping, bed and breakfast, carving and cattle-raising projects.

Crooked Tree Wildlife Sanctuary

Peace Corps Volunteer Jane Mackler, wife of Robert Mackler, helped the women of Crooked Tree Village to organize the Progressive Ladies Group.The group was formed in September, 1993, sponsored by BAS. In March of 1994 they received a grant from the United Nations Development Programme under "Partners in Development," to work on income-generating projects.

CTWS buffer villages include Biscayne, CrookedTree, Gardenia, Lemonal, and Maypen. BAS facilitated the development of a Cashew Cooperative with a Steering Committee composed of local, government, and NGO representatives. A processing plant was constructed and equipment purchased.

A feasibility study was conducted for a fish-farming project in the village of Crooked Tree, but it was deemed infeasible due to lack of appropriate quality and quantity of water.

BirdLife and BAS supported a major project to strengthen community support for the Crooked Tree Wildlife Sanctuary through identification of alternative livelihood activities. A series of training workshops included product development, business management, conflict resolution, environmental education, and stewardship. Activities included making jams and jellies, cuisine, table decorations and setting, arts and crafts, public relations, and communications.

One issue threatening the ecological integrity of the Crooked Tree Wildlife Sanctuary wetlands in 2008 was the increase in cattle farming adjacent to the lagoons. BAS addressed this issue by conducting training workshops with cattle farmers on best practices and developing a cartoon book that demonstrates these best practices.

BAS conducted a Wetland Summer Camp for teachers from Crooked Tree Wildlife Sanctuary buffer communities on "*Crooked Tree wetlands ga wi baak; as wi develop, mek wi kipit laik dat!*" (Crooked Tree wetlands supports us; as we develop, lets keep it intact!) The teachers were so enthusiastic that they shared their new knowledge at their Parent Teachers Association meeting in 2008.

St. Herman's Blue Hole National Park

The women of Armenia Village expressed a desire to develop a craft shop, so BAS set up an exchange with the women of Maya Center. In 2005 HIVOS funded a project that included support for the Armenia Women's Group, which resulted in the construction of a gift shop. The youth of the Amenia Village were trained in natural resources management and monitoring. They call themselves "the Guardians of St. Herman's Blue Hole and the Environment."

Olivia Carballo-Avilez

Armenia Village youth in training to be "Guardians of St. Herman's Blue Hole and the Environment."

Children from Armenia explore St. Herman's Blue Hole National Park

Olivia Carballo-Avilez

Outside lecture by Olivia Carballo-Avilez during training of Armenia Village youth.

Guanacaste National Park

Every year, the Belize Audubon Society is invited to the Ontario Christian Academy's Open Day. This event is held to showcase to the general public the different programs and activities the students have learnt at school. The principal also gives Belize Audubon Society the opportunity to showcase the work of Society for the visitors to the event. Ontario village is located on the Western Highway and is one of the several communities that are stakeholders of Guanacaste National Park and Tapir Mountain Nature Reserve.

Ontario Christian Academy's Open Day 2004

Half Moon Caye and Blue Hole Natural Monument

Stakeholder communities with fishermen who use Lighthouse Reef include Chunox, Copper Bank, and Sarteneja. Training workshops were held in the production of handicraft using conch shells, whelk shells and seeds to make earrings, pendants, bracelets, napkin holders, paper weights and other ornamental souvenirs. These souvenirs are sold in the BAS gift shops to complement the fishers' incomes (*top right*).

On June 24th and 25th a group of teachers from the Lighthouse Reef stakeholder communities were taken to Half Moon Caye to experience firsthand the value of this area to Belize and especially to their communities.They developed and delivered lessons on assigned topics for particular age groups.

Sponsored by the COMPACT programme and BAS, Lighthouse Reef stakeholder fishers took a field trip to Punta Allen in Quintana Roo, Mexico. Alternative fishing methods and an improved method for lobster harvesting using shades were demonstrated. The fishers were trained to become licensed tour guides and assisted BAS in monitoring of sea grasses, coral, conch, and lobster (*bottom right*).

Orlando Jimenez

Orlando Jimenez

John Arana

BAS produced a video entitled "The Past, Present, and future of fishing in the Lighthouse Reef Atoll/ *El Pasado, El Presente y El Futuro de la pesca de Lighthouse Reef Atoll* to document the state of fishing activities in Lighthouse Reef Atoll and its effect on fishermen stakeholders and their families. The video was taken on a tour of the stakeholder villages. Pre- and post-surveys were conducted among primary and high school students. The results indicated that the message that was strongest was to protect the fish now to ensure that there will be fish in the future.

John Arana

John Arana

(*top left*) BAS Community Liaison Officer Orlando Jimenez admires the arts and crafts in the village of Chunox (*bottom left*) Community members from Chunox and Copper Bank who participated in training to increase community capacity and product development.
(above) Earrings made in Chunox with seeds of the royal palm.

Friends of World Heritage provided funds in 2007-2010 for community empowerment through alternative livelihoods for conservation of the Belize World Heritage sites. A pilot project set up communal banks for Chunox, Copperbank and Sarteneja where community members became shareholders to provide small loans to their community.

BAS engaged the communities in an awareness campaign about the importance of protecting and using the natural resources in a sustainable manner. This included the door-to-door distribution of 1000 posters on the revised Fisheries' laws.

The awareness campaign culminated with 136 community members including teachers, fishermen and their families going on trips to Half Moon Caye and Blue Hole Natural Monument where they experienced first-hand the work of BAS. They learned about the history and importance of the sites and their unique resources, such as Red-footed Boobies, littoral forest, and turtle nesting sites.

The final product of the project was the publication of *"Partnering with Communities through Participatory Processes,"* a socio-economic report on the buffer communities for the nine protected areas BAS co-manages. For this effort, BAS worked with the Belize Red Cross, the Ministry of Rural Development, Village Councils and community-based organizations.

The Copperbank Impresio de Credito Comunal was inaugurated on December 9, 2009. (*top*) Losita Lee and Lucito Ayuso (*at the podium*) at Copperbank for the ribbon-cutting ceremony and cultural presentation (*bottom*).

BAS as Co-manager

Dominique Lizama and Anna D. Hoare

For the Belize Audubon Society co-management is two-fold: the Society has a formal co-management agreement with the Government of Belize, and secondly, partners with communities near protected areas.

In our years of creating partnerships with communities, Belize Audubon Society has had many successes, but there have also been shortfalls which have set the basis for future work with these communities and from which vital lessons have been learned. Many local communities who depend on these very resources for their livelihoods reject conservation efforts. Ongoing challenges include resentment towards the Society and protected areas, division within the communities between those who support and those who are against conservation, an existing great dependency on traditional use of natural resources, expansion of communities near protected areas, and the perception that communities do not benefit from protected areas whether directly or indirectly.

Economic hardships have made it more challenging for field staff members to enforce the laws against fishing, hunting, and plant extraction within their own communities and to confront people from outside the buffer communities who engage in illegal activities within the sites.

However, experience has demonstrated that communities can be empowered to participate in conservation efforts and still gain benefits while keeping the integrity of the biodiversity in the area intact. In partnering with these communities, BAS has worked towards positivity and has established Local Advisory Committees within the communities. Among the BAS permanent field staff, 85% come from the buffer communities and we have hired others on different occasions to assist in research and on site management duties. Community members have been involved in various trainings and meetings where concerns, laws, and projects have been discussed. Their expertise has also been sought for tours, catering, and hospitality for groups visiting Belize.

A great milestone for BAS was the development of memoranda of understanding between BAS and the Maya Center Women's Group and the Crooked Tree Village Council where each gets 10 percent of the protected area entrance fee.

We have realized the need for consultation with local people to help them understand the purpose of establishing protected areas and the benefits of conservation to their own livelihoods. So, the Society has implemented environmental education programs on birding, summer camps, and social marketing campaigns; construction of the craft shop in Maya Center; community banking in the northern communities; training in making handicrafts, tour guiding, fire management, better livestock management, and alternative livelihoods such as honey and cashew production.

Increased community capital translates into improved food security, stability, and health. It is imperative that BAS and communities understand that not all community livelihood projects will be successful, that communities need to build systems to create continuity in their livelihoods rather than depending on project funding which is time bound, and that management of natural resources requires a long-term commitment from communities.

This case study was presented by Dominique Lizama at the XIII Mesoamerican Society for Biology and Conservation Conference in October 2009 in Belize City.

Funding Agencies

None of this work would be possible without the streadfast support of the following funding agencies, which are in alphabetical order. BAS is grateful to every one of these organizations and apologizes for any inadvertent omissions.

AVINA Foundation

The AVINA Foundation supports individual leaders, rather than organizations. From 2003 through 2008 AVINA provided support for the BAS Executive Director, not only salary support, but additional support for programmes of the Society.

British High Commission

In late 1994 the British High Commission donated $5,000 to support the development of TMNR. The British High Commission was instrumental in securing sponsorship from a British organization for the publication of the second edition of *A to Z of Belizean Wildlife*. In 1996 they provided the materials needed for Raleigh International to build a private cabin at the CBWS.

Canada Fund

The Canada Fund provided the materials for a Community Resource Center for the Tapir Mountain Nature Reserve, which was constructed by Raleigh International.

Major Supporters	
2003-present	AVINA Foundation
1990-1994	Coastal Zone Management Project
1999-2006	Commission of the European Union
2005-2008	Darwin Initiative Programme (U.K.)
2001-present	Debt for Nature Swap (Governments of Belize and the United States)
2006-2010	Ford Motor Company
2007-2010	Friends of World Heritage
2000-present	Gordon Foundation
1998-2007	HIVOS
2003-2006	Homeland Foundation
2001-present	Jaguar Conservation Trust
2002-2004	Japan Official Development Assistance
1984-2002	John D. and Catherine T. MacArthur Foundation
1985-1991	Massachusetts Audubon Society
2000-present	National Audubon Society (U.S.)
2004-2006	National Fish and Wildlife (U.S.)
2004-present	Oak Foundation
2000-2005	Organization of American States
1998-present	Protected Areas Conservation Trust
1999-2003	Summit Foundation
2000-present	The Nature Conservancy
1994-present	United Nations Development Fund
2005-2007	United Nations Development Fund (COMPACT)
2007-2010	United Nations Education, Scientific and Cultural
1992-present	Wildlife Conservation Society
1999-2005	Wildlife Conservation Society (Wilson Foundation)
1985-1996	World Wildlife Fund-U.S.

Lighthouses at Half Moon Caye

Centro Agronomico Tropical de Investigacion y Enseñanza

Centro Agronomico Tropical de Investigacion y Enseñanza (CATIE) provided the funds for the purchase of the last privately-owned lots on Half Moon Caye in the mid-1970s. CATIE provided two grants, in 1979 and in 1984, for the development of Half Moon Caye Natural Monument. In 1977 CATIE provided a screen and tape recorder for the environmental education programme carried out in the schools. David Lovell was sent from CATIE in 1984 as a consultant for the drafting of the first management plans for BAS-managed protected areas.

Coastal Zone Management Project

The Coastal Zone Management Project (CZM) provided the first salary for an Environment Educator for the BAS staff and for the first warden at Half Moon Caye Natural Monument (HMCNM). Support has also been given for equipment for HMCNM. CZM has collaborated with BAS on exhibitions and other projects.

Commission of the European Union

The Actions in Favour of Tropical Forests Programme of the Commission of the European Union gave BAS a large grant in 1997 to support the three-year project, "Development of Revenue Generation Systems". Under this programme, entrance fees were instituted in most of the BAS-managed protected areas; BAS established a Marketing and Membership programme; tourism facilities were upgraded at CBWS, CTWS, BHNP, GNP, and HMCNM; and the management presence was enhanced at all protected areas through personnel recruitment and natural resource management training.

From 2000–2006 the European Union Commission supported the development of a co-management relationship with the buffer communities of the Cockscomb Basin and Crooked Tree Wildlife Sanctuaries.

Debt for Nature Swap

In 1999 the U.S. government passed a law entitled "Tropical Forest Conservation Act (TFCS)" which redirected money owed to the U.S. from developing countries into the protection of tropical forests. After a nine-month negotiation, an agreement was completed between the U.S., the Government of Belize, PACT, The Nature Conservancy, and three environmental Belizean NGOs: Programme for Belize, Toledo Institute for Development and Environment, and BAS. Under this agreement, the Belizean government agreed to repay their debt at reduced rates by annual grants to the NGOs. The result is that BAS receives BZ$ 160,000 per year until 2021 for protected area management. For 20 years, funds are deposited into a BAS endowment account that will build to BZ$ 3.4 million based upon interest earned at five percent.

HIVOS Funds for BAS Advocacy Programme

A Dutch NGO, the Humanitarian Institute for Cooperation with Developing Countries (HIVOS), supported the BAS Advocacy Programme from 1998 through 2007. The goal of the programme was to take action to ensure that policies set forth by the Government of Belize, especially with regard to land, air or water, are environmentally sound and take the best interest of the Belizean people into consideration. HIVOS also supported the development of the second BAS Strategic Plan 2002–2006.

The famous Guanacaste tree for which the Guanacaste National Park was named.

MacArthur Foundation

In October of 1984, the John D. and Catherine T. MacArthur Foundation of Chicago approved a grant of $20,000 U.S. yearly for three years to the Belize Audubon Society through the New York Zoological Society. For the second and third year of the grant period, BAS was to provide matching funds.

The grant was to be used for the management, development and, if necessary, the acquisition of parks and protected areas in Belize. The first project was the development of the Blue Hole National Park. Some of the money was spent on Guanacaste Park and a small sum was allocated to the proposed Hol Chan Marine Reserve near Ambergris Caye.

Additional grants have been received from the MacArthur Foundation for "Institutional Strengthening" (1990-1993), "Community Outreach and Protected Areas Development" (1993-1996), "Advocacy" (1996-1999), and "Education for Sustainability" (1999-2002).

BAS Archives

Mark Nolan of NARMAP presents Institutional Building grant to BAS, which is accepted by BAS President Therese Rath

Massachusetts Audubon Society

Beginning in 1985, the Massachusetts Audubon Society organized the Audubon Alliance, which was composed of several other Audubon Societies in the United States, to provide the first institutional development support to BAS (*for details see Chapter one*).

NARMAP

WWF-U.S. personnel administered and supported the Natural Resources Management Programme (NARMAP) funded largely by USAID. NARMAP made a number of contributions to BAS. In October, 1993, they sponsored the seminar entitled Economics of Natural Resources. Under its Institutional Building Component, NARMAP provided BAS with three new computer units in March 1994. In November BAS received a grant from NARMAP to upgrade infrastructure and develop revenue-generating activities at Crooked Tree Wildlife Sanctuary and Guanacaste and Blue Hole National Parks. The grant also facilitated bringing student groups to the area and conducting public awareness programmes. In 1996 NARMAP funds were granted to develop new facilities at CBWS including a kitchen and dining room, a bunkhouse accommodating 24 persons, composting toilets, and a solar electrical system for lights and showers.

National Audubon Society (U.S.)

The National Audubon Society (NAS), through Gordon Foundation funds, supported the development of a comprehensive plan for the CBWS. They also donated a radio communication system for the protected areas that BAS manages. NAS donated binoculars for all the BAS park personnel and backpacks with binoculars and other equipment for the Crooked Tree Junior Birding Club.

National Audubon Society, through the Gordon Foundation, donated a brand new Toyota vehicle for use at Cockscomb Basin Wildlife Sanctuary in 2006. The NAS also provided support for the social marketing campaign in Crooked Tree. In January 2009, Executive director Anna D. Hoare was invited for a two-week visit with National Audubon Society in the US. She visited the National Audubon Green building in New York, the Greenwich Nature Center, North Carolina Audubon Society, New York Audubon, and Connecticut Audubon to discuss possible collaborations.

(*from left*) Nicacio Coc, Bonifacio Cal, and Anna D. Hoare receive binoculars from Matthew Jeffrey of the National Audubon Society.

New pier on Half Moon Caye funded by PACT in 2007

The Nature Conservancy

The Nature Conservancy has provided support for the management of the Tapir Mountain Nature Reserve (TMNR) through the Maya Forest Small Grants Programme, which was funded by USAID. The grant also included funds for a comprehensive community outreach programme for TMNR.

The Nature Conservancy has partnered with BAS for development of public use plans for Lighthouse Reef and for the protection of spawning aggregation sites in Belize.

Oak Foundation

The Oak Foundation provided support for the work with the Lighthouse Reef stakeholder communities to involve fishers in management activities such as monitoring to build support for park enforcement and conservation efforts. Most importantly the Oak Foundation provided support for the long awaited Booby House that was started in August 2009 and inaugurated in May 14, 2010 and was to become one of the best warden's house in the country.

Protected Areas Conservation Trust

A grant from the Protected Areas Conservation Trust (PACT) supported BAS' Public Education Campaign begun in June 1998. The programme involved media spots aimed at educating the public about protected areas in Belize, with the intent that more Belizeans will visit these natural areas and develop a stronger appreciation for their country's natural resources. PACT has continued to support BAS management of protected areas the three year project entitled "Towards the Sustainability of BAS Managed Protected Areas" to complete activities at Half moon Caye and Cockscomb Basin Wildlife Sanctuary. The sum of $800,000.00 from 2006-2009 assisted BAS by providing for signage, boundary demarcation, training and equipment acquired for staff and major facility and infrastructure upgrades, such as a new pier at Half Moon Caye. PACT has supported special events such as Earth Day, Wetlands Day, and Turtle Week celebrations, as well as travel for BAS Staff to international meetings. PACT has been very supportive in emergencies, such as, BAS relief work after Hurricanes Keith, Iris, and Dean.

Judy Lumb

Baalum Naj cabin in the Cockscomb Basin Wildlife Sanctuary, that housed the camera trap studies and the work with cattle ranchers sponsored by the Wildlife Conservation Society.

Ramsar Convention Bureau

The Ramsar Convention Bureau has provided support for a programme of Conflict Resolution at the Crooked Tree Wildlife Sanctuary.

Summit Foundation

The Summit Fondation provided major support for the marine protected areas programme at BAS from 1999 until 2003 when the Summit Foundation had to cut their grants and could not fulfill their obligations. But they arranged for the Oak Foundation to take over those obligations.

United Nations Development Programme

The Global Environmental Facilities Small Grants Programme awarded a grant to BAS in 1994 to conduct a series of training workshops for protected areas personnel and to establish educational centers in the BAS-managed protected areas.

Weeden Foundation

The Weeden Foundation has contributed $200,000 to the BAS Endowment Fund.

Wild Wings Foundation

The Wild Wings Foundation has provided support for the Crooked Tree Wildlife Sanctuary since it was declared and continues to support the Sanctuary. They also provided institutional support for the management of protected areas.

Wildlife Conservation Society

Wildlife Conservation Society (WCS—formerly the New York Zoological Society) provided the funds for the jaguar study that led to the establishment of the Cockscomb Basin Wildlife Sanctuary and for the reintroduction of black howler monkeys into that Sanctuary.

Jaguar captured in a camera trap in CBWS (*Courtesy Wildlife Conservation Society*)

WCS, through their Wilson Fund, initiated a biodiversity monitoring project in the BAS-managed parks and created the position of Natural Resources Data Manager position at BAS. The project was in collaboration with Bruce Miller, administrator of the Belize Biodiversity Information System.

WCS supported jaguar research in the Cockscomb Basin Wildlife Sanctuary, including camera trap studies and work with cattle ranchers in the area. The Baalum Naj cabin in the Cockscomb Basin Wildlife Sanctuary was renovated for the camera trap studies by Bart Harmeson, and the work with cattle ranchers by Rebecca Foster.

World Wildlife Fund

The World Wildlife Fund of the United States (WWF-U.S.) first supported the environmental education efforts of the BAS. WWF-U.S. has supported the Cockscomb Basin Wildlife Sanctuary since it was established. A WWF-U.S. grant supported manatee and sea turtle research in the Gales Point area.

Many, Many Small Grants and Donations

BAS is grateful for the many small grants and donations that our members and friends have provided. It is this consistent and loyal support that has kept the Society working for a balance between people and the environment these forty years. Some of you contribute by your presence at BAS events thoughout the year and by participating in BAS fund-raising events. You are too numerous to mention by name, but be assured, BAS is very grateful to every one of you. Many thanks to one and all!

Fund-raising Campaigns

The campaign to gain corporate sponsorship was begun in 1990 when a letter was sent to Belize's foremost corporations. As a result of this and subsequent appeals, corporate sponsors have been very generous in their support of conservation through BAS.

The Friends of the Parks campaign was started in 1995, developed by VSO Volunteer Diana Houtzager. It allows members of BAS to directly support one or all of the BAS-managed parks.

The following list includes past and present corporate sponsors and Friends of the Parks. BAS is grateful to one and all!

Platinum Sponsors

Kim Aikman
Luis Alpuche
American Airlines
Belize Electricity Limited
Belize Global Travel Services Ltd.
Bowen & Bowen Ltd
Continental Airlines
Love FM
Pelican Beach Resort
Norman Pennill
Radisson Fort George Hotel
Tony Rath
Stephen L. Usher
Rene Villanueva
Denver Willson-Rymer
Terry Wright

Gold Sponsors

Atlantic Bank Ltd.
Belize Bank Ltd.
Immarbe
Krem Radio
Multi-Media Systems
Regent Insurance Co. Ltd.
Sydney Turton

Silver Sponsors

Alliance Bank
Belize Adventures
The Belize Glass Shop
Belize Telemedia Ltd.
Benny's Homecenter Ltd.
Emma Boiton
Tim Boys
Caligo Ventures Tour & Travel
Campbell Robert
Caribbean Tobacco
Caye Chapel Ventures
Citrus Products of Belize Ltd.
W. H. Courtenay & Co.
Coye & Associates
Esso Standard Oil S.A. Ltd.
F. Luciano Cuellar
Shaun Feinstien
Anthony Flynn
Island Expeditions Co.
The Guaardian
Audrey Matura-Tillett
Colin W. Medhurst
Pine Lumber Company Ltd.
Denver Willson Rymer
Margaret & Phil Schaeffer
Sittee River Wildlife Reserve

Bronze Sponsors

Advance Advertising Ideas
Alvaro Alpuche
G. Ayuso & Sons Ltd.
Belize Bird Rescue
Belize Marketing Board
Belize Sugar Industries Ltd.
Belize Tourist Board
Channel Broadcasting Corp.
Michael Estephan
FM 2000
James A.P. Fuller
Glenn D. Godfrey & Co
Colin Howell
Rufino Lin
Vildo Marin & Sons Ltd
Donald S. Morris
Lamanai Outpost Lodge
Placencia Investments
Michael Plowey
Sugar City Radio
Yalbac Ranch & Cattle Corp.

Friends of All Parks:

Belize Electric Company Ltd.

Friends of Two Parks:

Belize Timber Ltd.
Ted & Mary K. Berry
Steven R. Brechin
Gilbert Canton
Barbara & Steven Gaskin
Green Friends
David & Rosemary Griggs
Immarbe
Angelo Mouzouropoulos
Dan, Elle, & Dillon Plunkett
Summit High School
Tracy Taegar
Ed Zetina

Friends of One Park:

Neil Bird
Diane E. Blake
Gunter C. Bonrad
Butterfly World Ltd.
Robert W. Ehrig
Randolph A. Enriquez
Gary Galt
Judy Hill
Peter Hill
A & S Leuthner
Sarah A. Macrae
Todd & Mary Miller
Jim Roast
ors Specialists
Carol L. Mahoney
Modern Civil Engineeineeng
The Monkey House
NASIG
Reef and Rainforest Tours
P. B. Robinson
Stephen & Therese Showler
Robert A. Thomas
Gareth Welsh
Pat Welsh

Past and Present: Officers, Directors and Trustees

On the occasion of our 40th anniversary, we would like to record our thanks to the 114 members who have served on our Board of Directors, as Officers and Directors. As volunteers, they have helped tremendously to make BAS what it is today.

A Gallery of BAS Presidents

James A. Waight
(1969–1986)

Dr. Victor Gonzalez
(1986–1990)

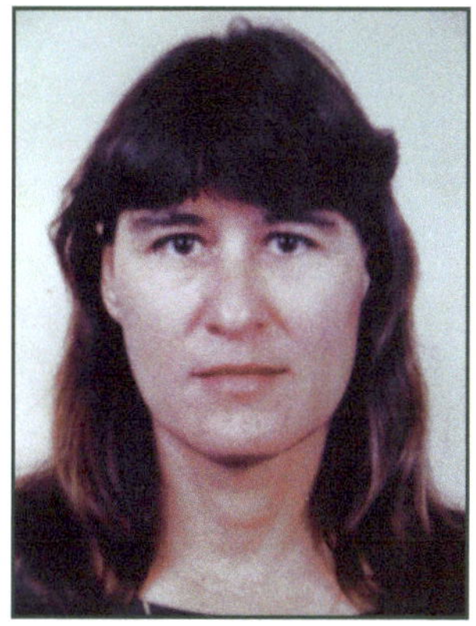

Janet Gibson
(1990–1991)

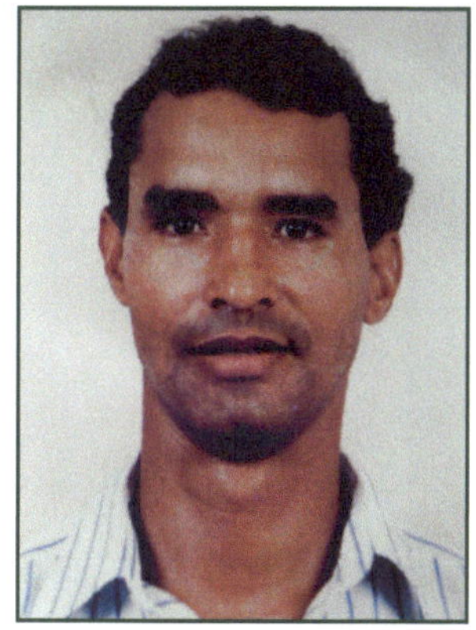

Philip Balderamos
(1991–1992)

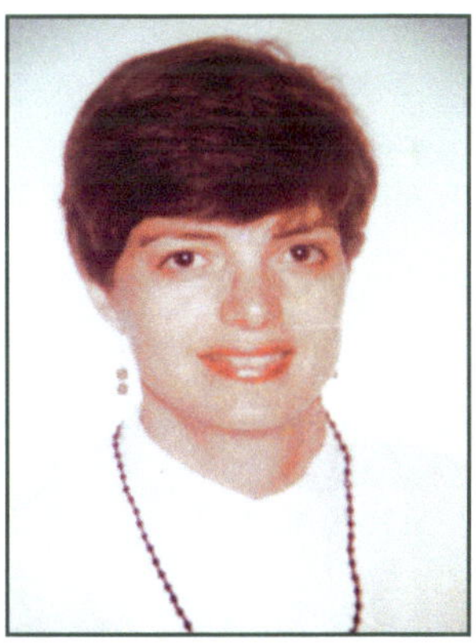

Therese Rath
(1992–1995)

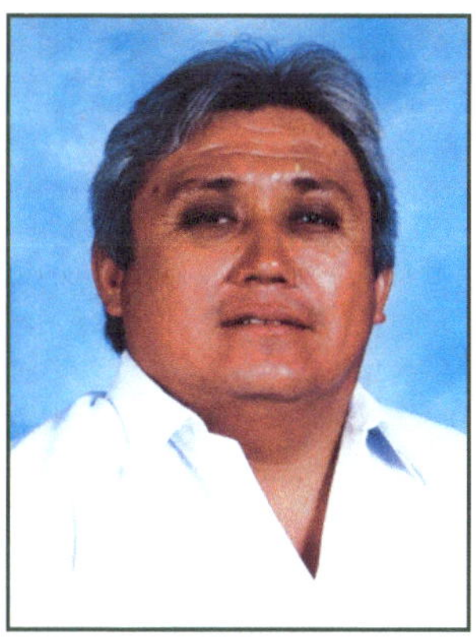

Jose Garcia
(1995–2000)

David Craig
(2000–2006)

Earl Green
(2006–present)

THANKS TO ALL OUR BAS BOARD MEMBERS

Lisel Alamilla
Nicole Auil
Oscar Ayuso
L.C. Balderamos
Philip Balderamos
John Balderamos
Jose Bautista
Amin Bedran
Chrisbert Berry
Wade Bevier
Barry Bowen
Rodolfo Burgos
Anselmo Castaneda
Marion Cayetano
Ronald Clark
Meg Craig
David Craig
L.J. Cuellar
James M. Currie
Flor David
Michael DeShield
L.E. Dieckman, S.J.
Ramiro Duran
Melissa Espat
Henry Fairweather
Anthony Flynn
Carlos Fuller
Anicasio Garcia
Jose (Pepe) Garcia
Janet Gibson
Ellajean Gillett
Dr. Victor Gongora
Kevin Gonzales
Dr. Victor Gonzalez
Albert S. Grant
Earl Green
Reynaldo Guerrero
Sylvano Guerrero
Magnus Halcrow
Norris Hall
William Hasse
Colonel Martin C. Hastings
Mike Heusner
Sister Joan Hickey, RSM
Anna Dominguez Hoare
Mark Hulse
Luz Hunter
C.L. Ingles
Michael Jenkins
Sister Marion Joseph, RSM
Lita Krohn
Homer Leslie
Louis Lindo
John Link
Jude Lizama
Joseph Loskot
Judy Lumb
D.E.G. Malone
Eric Marin
Magali Marin
Dr. Cardo Martinez
Sharon Matola
Shelly Matus
Carlos Matus
Shelley McCrary
Mark McField
Melanie McField
John McGarity
Barbara McLeod
Martin Meadows
Winston Miller
Beaveau B. Nalle
D.K. Ness
Lou Nicolait
Russell Norris
Humberto Paredes
Dr. Mary Pearl
Arlenie Perez
Therese Rath
Ernest Raymond
Lombardo Riverol
Dean Roches
Gil Rosado
Anna Rossington
Carlos Santos
Dr. Peter Schmidt
Patrick Scott, Sr.
Linda Searle
Jean Shaw
Wilmot Simmons
Michael Singh
Albert Staine
Norman Staines
Paul Szabo
Roger Taylor
Mario Teul
Sam Tillett
Ambrose Tillett
Lascelle Tillett
Henry Tromm
Edward Turton
J. Gonzalo Vasquez
Seymour Vernon
James A. Waight
Lydia Waight
Leo Weber, S.J.
Karen Weir
Dora Weyer
J. Charnock Wilson
Charles Woods, S.J.
W. Ford Young
Carlton Young

2006 BAS Board
(*from left*) Lascelle Tillett, Nicole Auil, Jude Lizama, Carlos Matus, Arlenie Perez, Carlos Santos, Mike Heusner, Carlton Young, Linda Searle, David Craig, Anna Rossington, Earl Green

2008 BAS Board
(*back row, from left*) Carlos Santos, David Craig, Lascelle Tillett, Carlton Young, Mark Hulse, Mark McField, Earl Green
(*front row, from left*) Carlos Matus, Anicasio Garcia, Linda Searle, Melissa Espat, Mario Teul

THANKS TO OUR BAS STAFF

Since 1985 it has been the paid Staff who have done the work of the Belize Audubon Society. This devoted group of people has served BAS in various positions for various lengths of time, often working overtime beyond the call of duty. It is their motivation and creativity that has made BAS the vibrant organization it is today. Thanks to all 148 of you!

Manciano Acal
Vincente Acal
Nikisha Ack
Amanda Acosta
Eugenio Ah
Valdemar Andrade
Christine Anthony
John Arrana
Roger Arana
Elisa Ariola
Magda Avery
Lucito Ayuso
Diane Bainton
Rita Baizar
Tyrone Bell
Brijilio Bolon
Mario Bolon
Donna Bull
Rudy Burgos
Bonifacio Cal
Elias Cantun
Gaspar Carballo
Olivia Carballo-Avilez
Cecy Castillo
Ted Castillo
Gillion Castro
Nellie Catzim
Celso Cawich
John Cayetano
Derric Chan
Daniel Chan
Ivis Chan
Jamie Lee Chanona
Venancio Cho
Dareece Chuc
Angel Chun
Juan Chun
Nicacio Coc
Arvin Coc
Francisco Coc
Ian Courtenay
Micky Craig
Vincent Crawford
Clara Cuellar
Veronica Dacoff
Angelino Dominguez
Transito Escalante
Fernando Esquivel
Kirah Foreman
Dirk Francisco
Marco Tulio Franco
Kenneth Gale
Hampton Gamboa
Erleon Garbutt
Albert Garbutt
Kent Garcia
Marceluz Geban
Allen Genus
Dolores Godfrey
Kendra Griffith
Marciano Guiterrez
Said Gutierrez
Antonio Gutierrez
Colleen Hamilton
Christopher Hecker
Derick Hendy
Dudley Heredia
Nimmi Herrera
Sergio Hoare
Anna Dominguez Hoare
Transito Jimenez
Ronnie Jimenez
Orlando Jimenez
Rennie Jones
Jairo Juchim
Marco Leal
Felicia Leal
Losita Lee
Yesenia Leiva
Albart Liconia
Michael Link
Dominique Lizama
Maria Luna
Soila Maheia
Israel Manzanero, Jr.
Israel Manzanero, Sr.
Giovanni Martinez
Leanne McCarthy
John Moguel
Mark Moody
Diane Wade Moore
Heron Moreno
Khaidia Mossiah
June Neal
Delia Noble
Luis Novelo
Marcelo Pau
Pedro N. Perez, Jr.
Neysi Pickwoad
John Pinelo
Sonia Pinto
Pedro Pixabaj
Ignacio Pop
Emiliano Pop
Ramona Pop
Agapita Pop
Roberto Pott
Sharon Ramclam
Maria Ramirez
Kanne Requena
Uriah Rhaburn
Ronald Richards
Julianne Robinson
Harvey Rosado
Natalie Rosado
Anna Rossington
Osmany Salas
Antonio Samayoa
Josiah Santos
Larry Santos
Ernesto Saqui
Pio Saqui
Patrick Scott, Jr.
Antonio Shal
Celso Sho
Emiliano Sho
Mark Smith
Michael Somerville
Stephen Swasey
Daryll Teo
Donald Tillett
Steve Tillett
Nancy Tillett
Darwin Tun
Frederico Tush
Andrew Tush
Oliver Twist
Mauricio Uk
Virginia Vasquez
Karen M. Vernon
Sylvestre Villanueva
Jason Wade
Diane Wade-Moore
Dennis Williams
Carol Williams
Tanya Williams-Thompson
Shane Young

BAS Hard-working Field Staff

Cockscomb Basin Wildlife Sanctuary wardens in 2003: (*back row*) World Gap Challenge Volunteers Bob MacKintosh and Olly Bates; (*from left*) Federico Tush, Abraham Coc, Mario Bolon, Brijilio Bolon, Bonifacio Cal, Ignacio Pop, Emiliano Sho, Ernesto Saqui, Nicacio Coc

Crooked Tree Wildlife Sanctuary wardens in 2003: (*from left*) Steve Tillett, Rennie Jones, Derick Hendy, and Donald Tillett

Summary and Conclusions

BAS was formed by a group of 55 enthusiastic and energetic conservationists on February 6, 1969, as a chapter of the Florida Audubon Society, and became an independent Society in 1973. For the past forty years, the Belize Audubon Society (BAS) has helped to make Belize a model of a developing country with an environmental consciousness. For the first fifteen years BAS was Belize's only environmental organization. Through a close working relationship with the Government of Belize, BAS worked to preserve the country's precious natural resources for the generations to come.

BAS was consulted on all proposed development projects and warned against those that would be damaging to the environment. The Society facilitated the early passage of legislation for the protection of wildlife and establishment of protected areas. The Society proposed areas that should be protected and lobbied until wildlife sanctuaries, natural monuments, nature reserves, and national parks were declared. BAS led the management of these protected areas. Indeed, Belize's current enviable position as a premier ecotourism destination is largely due to the early work of BAS.

BAS has developed in three phases. During the first fifteen years the work of the Society was done solely by volunteers using donated materials. By the end of 1984 some funds were obtained to support management of protected areas and establish an office with professional staff. The next five years were a volatile time with inconsistent funding, changes in the Government of Belize, and enormous developmental issues for the country. By 1990 the situation had stabilized and the professional phase began in earnest. The BAS staff has continued to increase in number and professional skill ever since.

Since the very beginning the Society has sat upon the same three-legged programmatic stool, the legs of which are: Advocacy, Environmental Education and Management of Protected Areas. Each conservation issue, each environmental education project, and each protected area is a story in itself.

BAS manages nine of Belize's public protected areas: Actun Tunichil Muknal, Blue Hole, Half Moon Caye, and Victoria Peak Natural Monuments; Cockscomb Basin and Crooked Tree Wildlife Sanctuaries; Guanacaste and St. Herman's Blue Hole National Parks; and Tapir Mountain Nature Reserve.

From the reintroduction of black howler monkeys into the Cockscomb Basin to monitoring of the spawning aggregation sites for Nassau groupers, BAS has been actively involved in research in biodiversity. Biodiversity monitoring and other research results are integral to the management of BAS-managed protected areas and the other conservation work of the Society.

Over the past 20 years BAS has given and received a number of awards. The James A. Waight Award for Conservation has been given in honour of the Society's first President since 1987. The profiles of the recipients provide a snapshot of Belizean environmental heroes.

BAS has accomplished all this with the help of its partners who have provided financial support and joined with us in the work of preserving Belize for a bright future. Belize's other environmental organizations, stakeholder communities buffering BAS-managed protected areas, international and local funding agencies, volunteer organizations, and corporate sponsors have all been our partners in conservation.

Index

Symbols

12 Fort Street Office 35
25th anniversary 39
30th anniversary 15, 40, 131
40th anniversary 14, 41-8

A

Ack, Heraldo 135
Ack, Martin 135
Ack, Nikisha 33, 221
Ackers, Ruth 190
Acosta, Amanda Burgos x, 13, 14, 76
Acosta, Charles 197
Actun Tunichil Muknal Natural Monument (ATMNM) 26, 123, 148-9, 170, 236, 266
Administrative Staff 11-4
Advocacy 7, 29, 32, 51-80, 87, 107, 119, 255
Advocacy Coordinator 13
Advocacy Technical Committee 68
AGM 7, 9-10, 18-24, 35, 139, 145, 157, 164, 222-4
Agriculture 65
Agriculture Show 93
Aguilar Revelo, Lorena 85
Ah, Eugenio 88, 136, 223
Alamilla, Lisel 18, 111, 210
Yanosky, Alberto 232
Alcorneopsis floribunda (*Euphorbiaceae*) 192
Alma Caribeña 48
Alsophila salvinii (*Cyatheaceae*) 192
Altun Ha 3, 6, 108, 174
Alvarado, Emil 48
Ambergris Caye 52, 55, 76, 207, 228, 255
American Bird Conservancy 111
Amigos de la Naturaleza 7
Amphibian and Biodiversity Expedition 190-1
ANDA 61, 65, 67, 68, 229
Andrade, Valdemar 13, 14, 18, 65, 83, 142, 184, 213, 224, 239
Angelus Press Ltd 110, 111
Anthony, Christine 13, 19
Annual General Meeting (AGM) 7, 9-10, 18-24, 35, 139, 145, 157, 164, 222-4
APAMO 63, 154, 229
Aquaculture 107
Ara Macao Project 75
Arana, John 30, 250
Arana, Roger 195
Archaeological Commissioner 61
Archeology, Institute of (IoA) 148-9
Ariola, Elisa 30
Armenia 247
Armenia Village Women's Group 235
Arnold, Lascelle 125
Association of National Development Agencies (ANDA) 61, 65, 67, 68, 229
Association of Protected Areas Management Organizations (APAMO) 63, 154, 229
Atkinson, Richard Charles 205, 226
Atlantic Bank, Ltd. 41, 48, 260
Atlantic Industries Corporation 59
ATMNM 26, 123, 148-9, 170, 236, 266
Auil, Nicole 20, 22, 31, 224, 263
Avery, Magda 27, 28
AVINA 117, 128, 243, 253
Awards 205-226
Azmax Filming Company 119
Azueta, James 195

B

BACONGO 61, 65, 229
BAHA 70
Baird, James 11
Baird's tapir 51
Baker, Gordon 165
Balderamos Garcia, Hon. Dolores 216
Balderamos, John 2, 18, 19, 20
Balderamos, Leo 2, 4
Balderamos, Philip viii, 5, 18, 27, 54, 171, 179, 216, 226, 261
Baldwin, D. H. (Don) 3, 108
Baltaupht Travels 119
Baraff, David 238
Barborak, James 57
Barmon, W. 2, 5
Barnette, Adrian 184
Barry, Johannah 27
Bates, Olly 265
Bayly, Dr. Nick 88
BCES 227
BCVI 104
Bedran, Amin 18
Belisle, Lindsay 18, 35, 36
Belize Agricultural Health Authority (BAHA) 70
Belize Agro-producers Organization 68
Belize Alliance of Conservation Non-government Organizations (BACONGO) 229
Belizean North-East Biological Corridor 88
Belizean Rainforest 109
Belize Association of Conservation NGOs 61
Belize Barrier Reef 55
Belize Biltmore Plaza 23
Belize Biltmore Plaza Hotel 48
Belize BirdFest 175
Belize Brewing Company 125
Belize Center for Environmental Studies (BCES) 227
Belize City Counci 41
Belize City Garden Club 40
Belie Coast Guard 56, 224, 226, 231
Belize Coalition to Save Our National Heritage 78
Belize Community Service Alliance 213
Belize Council for the Visually Impaired 104
Belize Dance Company 39, 233
Belize Electricity Ltd 67, 126
Belize Enterprise for Sustainable Technology (BEST) 27, 230, 243
Belize Family Life Association 106, 226
Belize Fishermen's Cooperatives 230
Belize National Herbarium 192
Belize Ports Authority 35, 56
Belize Red Cross 41, 154, 245, 251
Belize Telemedia Ltd. 48
Belize Tourism Board 48, 226
Belize Tourism Expo (BETEX) 132
Belize Tourism Industry Association (BTIA) 62, 63, 162
Belize Water Authority 125
Belize Youth Conservation Corps (BYCC) 104, 238

Belize Zoo and Tropical Education Centre 40, 111, 228
Belmopan Comprehensive School 39
Ben's Bluff 158
BEST 27, 230, 243
Best Practices Booklet 133
best practices for cattle farmers 74
BETEX 132
Beveridge, James 12, 112, 181
Big Creek 56, 69
Borella, Bill 89, 90, 126, 141
Biltmore Plaza Hotel, Belize 20, 23, 48
biodiversity 32, 49, 77, 92, 93, 114, 117, 128, 133, 150, 171-204, 220, 252, 259, 266
biodiversity monitoring 180-1, 199-204
Birdathon, Neotropical 176
Bird Checklists 108
BirdFest International 175
Bird Festival at CTWS 102
birding 5
BirdLife International 107, 111, 232-5
Birds without Borders 230
Biscayne Primary School 99
Bishop, John 29
Black-bellied Tree-Duck 53
Black Creek 71
Bladen Branch Nature Reserve 134
Bliss Institute 93
Blue Hole National Park (SHBHNP) 10, 140-145, 255
Blue Hole Natural Monument 15, 55, 131-133
Blue Morpho (*Morpho peleides*) 44
Board of Directors 17, 52
Board of Trustees 17
boardwalk 153
Bocawina Falls 179
Boles, Dr. Ed 183, 184, 220-1, 226, 285
Bolon, Alejandro 190
Bolon, Brijilio 87, 265
Bolon, Mario 30, 32, 265
books 109-111
Boomsma, Tineke 184
Borland, Commandant John 226
Bowen, Barry 40, 224
BRC Printing 109, 110, 112
Brewer, Steven W. 192
Briceño, Hon. John 18, 25, 35, 36, 40, 66, 131, 142, 232
British Executive Service Overseas 29
British Forces Belize 188, 241
British High Commission 104, 109, 153, 164, 165, 240, 253
brochures 112-113
BTIA 62, 63, 162
bullet tree (*Bucida buceras*) 151
Bulletin, BAS 7, 9, 52, 81, 83, 107, 122
Burgos, Rudi 27, 28, 106, 179
Burton, W. B. 184
BYCC 104, 238

C

Cacao Farmers 245
Cahal Pech 28, 149
Cal, Bonifacio 31, 33, 163, 191, 257, 265
Callahan, Laura 136
Camping Trail 142
Camps, Environmental 104-106
Canada Fund 147. 253
Canadian Audubon Society 108, 150
Cangrejo Caye 62
Carballo-Avilez, Olivia 48, 89, 92, 99-103, 105, 121, 138, 163, 247
Carnival Cruise Terminal 75
Carr III, Archie 57, 89, 159
Carter, Ann 2, 5
Carter, Bolon 5
Carter, Jacque 110
Cashew Fest 93
Castenada, Anselmo 18
Castillo, Beverly 100
Castillo, Rudoph 208
Castillo, Ted 13
Castro, Gillian 33
Castro, Hon. Edmund 71, 100
CATIE 57, 124, 125, 126, 140, 159, 254
cattails (*Typha domingensis*) 182
Cattle in Crooked Tree 73-74
Catzim, Nellie 61, 213
Causeway 151
CAVU 118, 119
Caye Caulker 52, 112, 115, 134, 174, 212
Cayo Parks 107, 134-149
CBWS 10, 20, 57, 158-168, 184, 242, 243, 244-5, 258
Center for Park Management 143
Center for Tropical Conservation of Philadelphia 84
Centro Agronomico Tropical de Investigacion y Enseñanza (CATIE) 57, 124, 125, 126, 140, 159, 254
Chaa Creek Resort 28, 30
Chavannes, Gerald 115
Chalillo Dam (Macal Upper River Hydro-facility) 67-8
Chamaedorea costaricana (*Arecaceae*) 192
Chan, Ivis 31, 131, 181, 190-1, 194
Channel Five 79
Chanona, Jamie Lee 14, 197
Chanona, Janelle 218-9, 226
Chau Hiix Mayan Site 154
Cherak's Farm 69
Chief Agricultural Officer 61
Chief Broadcasting Officer 52
Chief Environmental Officer 61
Chief Forest Officer 52, 61
Chief Game Warden 52
Chief Hydrologist 61
Chief Veterinary Officer 57
Cho, Julian 18, 210, 226
Cho, Maggie 210
Cho, Sebastian 190
Choc, Juan 135
Christmas Bird Counts 171-173
CITES Convention 51
Clark, Ronald 1, 2, 4, 51
Clewell, Dr. Andre 7, 82
Climate Change 66, 79, 87, 96, 107, 221
Coastal Zone Management Project 13, 56, 83, 84, 126, 198, 207, 254
Coast Guard, Belize 56, 224, 226, 231
Coc, Abraham 265
Coc, Andrea 45
Coc, Arvin 41
Coc, Nicacio 24, 89, 163, 223, 257
Coc, Venancio 33
Coca Cola 58-9
Cockscomb Basin Wildlife Sanctuary (CBWS) 10, 20, 57, 158-168, 184-9, 242, 243, 244-5, 258

Code of Conduct 18
Collado, Pablo 48, 233
Colon, Christina 184
colouring books 114
Columbia Forest Reserve 59, 229
Columbia Forest Wildlife Refuge 134
co-management 107
Comic Books 114
Commissioner of Lands 61
Community Baboon Sanctuary 18, 228
Community Liaison Officer 14
Community Management of Protected Areas for Conservation (COMPACT) 249
COMPACT 249
conch 130, 198
Convention on International Trade of Endangered Species of Flora and Fauna (CITIES) 51
Copper Bank 249
Coral Reef 56, 87, 109, 199-204, 257
Costa Rican Embassy 41
cotton cultivation 57
Courtney, Ian 231
Cowo, Manuel 73
Cox, Siobhan 141
Craig, Alice M. (Meg) x, 1, 2, 18, 82, 125, 171, 217, 225, 226, 297
Craig, David 1, 66, 88, 125, 210, 224, 238, 261
Craig, Elise A. 8
Craig, Walter (Micky) 1, 112, 159, 171, 206, 237
Crane, Kent B. 57
Crane Game Ranch 9, 58
Creamer, Steven 136
Crested Eagle (*Morphnus guianensis*) 190
Crooked Tree Government School 101
Crooked Tree Lagoon 69
Crooked Tree Wildlife Sanctuary (CTWS) 7, 14, 25, 28, 33, 34, 38, 49, 57, 69-72, 92, 150-157, 182-3, 243, 246-7, 265
Cross, William E. 177
Crown Reserve Bird Sanctuaries 134
Cruz, Yassiri 16
Crystal Cave 106
CTWS 7, 57, 71, 92, 150-157, 182-3, 243, 246-7
Cuevas, Eloy 214
Curriculum Development 109

D

Dakan, Robert 18
Damselflies 147
Dangriga Youth Group 41
Dasyatidae 202
Davis, Katharine 141
Debt for Nature Swap 255
DelMargo, Ellen Waight 218, 219, 220, 221
DelMargo, Richard 221
DEMA 116, 132
Department of Environment (DOE) 56
Dependence on Grants 34
Diaz, Perry 46
Dieckman, S.J., Fr. Len 2. 4
Dietrich-Ward, Svea 40, 146
Director of Geology and Petroleum 61
Diving Equipment and Marketing Association (DEMA) 116, 132
DOE 68
Dolphin Theme Park 62
Dorothy Menzies Childcare Centre 130
Drowned Caye 75
Topalian, Dr. Teny 18
Drypetes guatemalensis (*Euphorbiaceae*) 192
Dubois, R. 57
Dusky Antbird Trail 142

E

Earth Day Celebrations 15, 95-98
Earth Summit 76
Eco-Fashion Design 48
Eco-Fest 39
ecosystems 171
Eden Conservancy 141, 235
Education Kits 90
Education Programme 7, 26, 81-122
Edwards, Scott 143
EIA 58
Eisenmann, Eugene 9
El Colegio de la Frontera Sur (ECOSUR) 182
Elliot, Philip 184
Embassy of Mexico in Belize 41, 42, 48, 98
Emergency Management Mechanism 68
Emmons, Katherine M. 184
Endowment Fund 17
entrance fees 33
Environmental Agenda 65-66
Environmental Conservation Carnivals 84
Environmental Education Programme 7, 26, 81-122
Environmental Impact Assessment 57
Environmental Leadership Camp 104
Environmental Youth Club 241
Espat, Luke 71
Espat, Melissa 263
Esquivel, Kathy 39
Esso Standard Oil 56
European Union, Commission of the 243, 254

F

Fairweather, Henry 1, 2, 4, 51, 210, 226
Family Life Association 226
Farmer-to-Farmer Programme 182
Faulkner, Dr. Otto 135
Feathers 'N Fins 93
field staff 14
field trips 5-6
Figueroa, Omar 178
filming 118
Finance Committee 17, 34
Financial Manual 34
Financial Sustainability 34
Financial Trials 33-34
Findley, Greg 143
fish kill in Crooked Tree 70
Fisheries Administrator 61
Fishermen's Cooperatives 230
Fisk, Emily 152
Flores, Titman 94
Florida Audubon Society 7, 134, 266
Flowers Bank 246
Flying Carpet Tours 108
Flynn, Tony 224
Fonseca, Hon. Ralph 18

Forest Department 26, 41
Foster, Rebecca 184, 187, 259
founders of BAS 2-5
Francisco, Dirk iv, x, 14, 38, 42, 93, 94, 99, 112, 171, 172, 179, 223
Freese, Curtis 11
Freshwater Creek 85
Friends of Nature 214
Friends of the Earth 58
Frigatebird, Magnificent 8, 108, 175, 177
Fuller, Kathryn 162

G

Game Ranch Project 57
Gap Challenge, World 141, 242
Garbutt, Erleon 33
Garcia, Anicasio 22
Garcia, Emmerson 181
Garcia, Hon. Russell 69
Garcia, Jose (Pepe) 18, 22, 25, 261
Garcia, Kent 31, 88
Garcia, Luis 213
Gardening Club 103
Garel, Rowan 96
Geban, Marceluz 132
GEF SGP 85
Gender and Development 85
Gentle, Lionel 27
Gestring, Bonnie 136
Gibson, David 162
Gibson, Janet 18, 27, 109, 114, 162, 195, 205, 207, 225, 226, 241, 261
gift shop at BAS 14, 35
Gillett, Katie 92
Gladden Spit 214
Glovers Reef 82
GNP 10, 17, 135-139, 238, 248
GOB 4, 7, 9, 25-6, 33, 34, 53-5, 57-8, 62-3, 65, 77-8, 123-4, 126, 134, 135, 146, 150, 159, 204, 208, 211, 229, 235, 237, 252, 255, 266
Godfrey, Dolores 12, 27, 28, 162, 207
Godfrey, Glenn 162
Golden Citizens 41
Golden-crowned Warbler 142
Goldman Environmental Prize 225
Gomez, Camila 88
Gonzalez, Dr. Victor 27, 162, 237, 261
Goodwin, Zoe 152, 190, 191
Goose, Canada 179
Goshen College 82
Government of Belize 4, 7, 9, 25-6, 33, 34, 53-5, 57-8, 62-3, 65, 77-8, 123-4, 126, 134, 135, 146, 150, 159, 204, 208, 211, 229, 235, 237, 252, 255, 266
Graham, Rachel T. 202
Grajal, Dr. Alejandro 18, 19, 20, 21, 24, 48, 111, 164,
Grant, Albert S. 1, 2, 4, 51
Grant, Joy 162
Green, Earl ix, x, 18, 216, 218, 224, 261
Green, Leroy 219
Green Art and Flower Show 40
Green Initiative 113
Green Reef-Belize 228
Green Sea Turtle (*Chelonia mydas*) 55
Green Wave Campaign 118
Griffith, Kendra 14
Guanacaste National Park 10, 17, 135-139, 238, 248
Guanacaste tree 139
Guardians of St. Herman's Blue Hole and the Environment 247
Guatemalan Embassy 41
Guerra, Madeleine 221
Guerrero, Reynaldo 18
Gutierrez, Said 187

H

Habeck, Robert 135
Habitat Map 197
Haematoxylum campechianum 151
Halcrow, Magnus 1, 2, 3, 7, 82, 135
Half Moon Caye Natural Monument (HMCNM) 7, 57, 124-130, 238, 249-51, 257
Hall, Wayne 179
Halle, Mark 57
Hamilton, Colleen 14
handicraft 249
Hanson, George 130
Happy Hollow Zoo 235
Harmesen, Bart 186-7
Harpy Eagle 111
Hasse, Bill 4
Hasse, Diane 107
Hastings, Colonel Martin C. 1, 2, 3, 83
Hawk-Eagle, Black and White 179
hawksbill turtle 199
Hecker, Christopher 14, 171
Hendy, Derick 30, 32, 88, 103, 265
Heredia, Dudley 43
Heredia, Lionel (Chocolate) 134, 212, 226
Herrera, Nimmi 61, 88
Heusner, Mike 4, 18, 19, 20, 24, 42, 63, 263
Higginbottom, Paul 187
Hill, Peter 2, 5, 7, 82
History Channel 119
HIVOS 30, 65, 255
HMCNM 7, 57, 124-130, 238, 249-51, 257
Hoare, Anna Dominguez 13, 38, 219, 252, 256, 257
Hoare, Sergio 197, 223
Hol Chan Marine Reserve 10, 208, 225, 255
Hope, Kathleen 2, 5
Horwich, Dr. Rob 18, 109, 188
Housing and Planning Officer 61
Houtzager, Diana 260
Hulse, Mark 19, 24, 216, 263
Humanitarian Institute for Cooperation with Developing Countries (HIVOS) 30, 65, 255
Hunkeler, Pierre 57
Hunter, Luz 27
Hurricane Dean 154
Hurricane Greta 135
Hurricane Iris 245
Hurricane Ivan 236
Hurricane Keith 137, 175
Hurricane Mitch 152
hurricane relief 154
Hyde, Deidre 164
Hyde, James V. 25, 123, 159, 208, 226
Hydrological Monitoring 68

I

ICTF 228
Iguana iguana rhinolopha 177
iguanas 177
Immigration Department 56
Important Bird Areas 173-174, 204, 232
Independence, BAS 7
International Conservation Relief 146
International Expeditions 108
International Tropical Conservation Foundation (ICTF) 228
International Whaling Commission 55
IoA 148-9
Island Expeditions 236
IUCN 27, 30, 237

J

Jabiru 52, 178
Northern Jacana 174
Jacanas 174
Jacobs, Jeremy F. 184
Jaguar Conservation Fund 164
Jaguar Conservation Trust 235
Jaguar Creek 105, 235
Jaguar Preserve 158
jaguars 90, 105,115, 116, 211, 244, 245
jaguar research 184, 185-7, 258-9
James, Dr. Douglas 87
James A. Waight Conservation Award 15, 205-221, 226
Jex, John 152
Jimbo Roo, 218
Jimenez, Orlando 21, 32, 243, 246, 249-50
Jimenez, Transito 33, 105
Jones, Lee 172, 179, 181
Jones, Rennie 32, 103, 152, 180, 182, 265
Juan, Hon. Eduardo 25
Junior Bird Club 103

K

Kangas, Patrick 184
Keel-billed Motmot (*Electron carinatum*) 190
Keel-billed Toucan 108
KEEP BELIZE GREEN, GO GREEN - IT MAKES ENE 97
Klemm, Cyrille de 57
Knight of the Golden Ark 225
Konecny, Michael 184
Koontz, Fred 188-9
KREM 79

L

La Bomba 246
Lamanai Room Declaration 62
Lamanai Room Declaration II 63-64
Land Alliance for National Development (LAND) 68
Land Policy Reform 68
Lauer, Matt 119
Leal, Felicia 32, 155
Leal, Roberto 103
Leberman, Robert C. 108
Lee, Losita x, 119, 132, 199, 251
Leiva, Yesenia 19
Le Levant 56
Lemonal 99, 246
Leslie, Robert 208
Lewis, Senator Conrad 162
Library 83
Liconia, Albart 31, 194
LightHawk 239
Lighthouse Reef 56, 199-204, 257
Limits of Acceptable Change 132, 196
Lindo, H.E. Dean 10, 59, 124, 135, 140, 160, 208, 226
Lindo, Louis 1, 2, 4, 7, 51, 82
Lineated Woodpecker 142
Link, John 27, 28
Link, Michael 180
lionfish 201
live rock exportation 62
Lizama, Dominique 1, 24, 30, 38, 48, 194, 222, 252
Lizama, Jude 20, 22, 263
lobster 62, 200
Local Advisory Committees 243, 252
loggerhead turtles 196, 199, 201
logging 59-60
logwood (*Haematoxylum campechianum*) 73, 151
Lolette 125
Longacre Expeditions 238
Lopez, Mayor Simeon 41
Love FM 41, 48, 79, 260
Lovell, David 57, 126, 152, 254
Luckner, David 56
Lyon, Jonathan 109, 184

M

Macal Upper River Hydro-facility (Chalillo Dam) 67-8
MacArthur Foundation 85, 90, 135, 140, 253, 255
MacFarland, Dr. Craig 124
MacKintosh, Bob 265
Mackler, Jane 246
Mackler, Robert 140
Maheia, Wil 179
manatee grass (*Syringodium filiforme*) 199
manatee 52, 93
mangroves 93
Manzanero, Jr., Israel 88, 105, 121, 180, 194
Manzanero, Sr., Israel 88, 105, 106, 141, 181, 223
marimba 20, 23, 164
Marin, Ameera 48
Marin, Francis 39
Marin, Hon. Florencio 136, 162, 208, 225, 226, 229
Marin, Magali 18
Marin, Silverio Y. 187
Marine Biologist 14
Marine Forum 93, 195
marine protected areas 14, 76, 80, 107, 119, 174, 179, 225
marine research 195-204
marine resources, protecting 55-58, 107, 133
Marine Welding 231
Mark Phillips Jazz Trio 43
Marlin, Jake 184
Marlin, Kelly 184
Martin, Capt. Ted 109, 241
Marrow, Greg 110
Marsh, Laura 184
Martinez, Giovanni 33
Mason, C. Russell (Russ) 1, 7
Massachusetts Audubon Society 11, 59, 111, 228, 253, 256

Matola, Sharon 162, 228
Matus, Carlos 22, 263
Matus, Shelly 4, 134
Maya Center 226
Maya Center Women's Group 24, 162, 164, 168, 244-5, 247, 252
Maya Mopan 23, 90, 138, 242, 244-5
Maya Walk Tours 148
McCarthy, Leanne 14
McCarthy, Timothy J. 112, 184
McCoy, Dr. C. J. 115
McCrary, Shelley 27-8
McField, Mark 263
McField, Melanie 63, 78, 195
McRae, Ellen 109
McSweaney, Andrea 39
Media Campaigns 79
Meekin, Catherine 83, 182, 240
Meekin, Tom 182, 240
Meerman, Jan 136, 179, 184, 228
Mega-Developments 75
membership 7, 15-17
Membership Convention 15-17
Membership Coordinator 14
Meso-American Society for Biology and Conservation 117
Migratory Birds 88
Miller, Bruce 69, 108, 136, 172, 184, 259
Miller, Carolyn 69, 108, 111, 136, 184
Miller, G. Winston 55, 209, 226
Miller, Matthew 135, 136
Miller, Winston 1, 2
Minister of Agriculture and Fisheries 69
Minister of Natural Resources 57
Ministry of Rural Development 251
Mitch Caye 239
Model, Allan 103
monkeys, black howler 105, 109, 184, 188-9, 206, 228, 258, 266
Monkey River Oil Refinery 77
Monkey Translocation 188-9
Moore, Alan 160
mooring buoys 128, 130
Moralez, Sarita 48
Morell, Merilio 57
Moreno, Excellency Manuel Lopez 41, 42
Moreno, Heron 14, 88, 224, 242
Mossiah, Khaidia 24, 30, 32, 48, 132
Motmot, Blue-crowned 181
Motmot, Keel-billed 179, 181, 190
Motor Vessel Transfer 56
Mountain Cow Cave 143
Moya, Mayor Zenaida 41
Muscovy Duck 53

N

NARMAP 18, 84, 85, 164, 229, 256
Nassau Grouper 110, 198
National Audubon Society (U.S.) 18, 103, 112, 114, 150, 171, 253, 256, 257
National Conservation Strategy 57
National Environmental Appraisal Committee (NEAC) 61, 67, 68, 75, 80, 229
National Geographic 118
National Institute of Culture and History (NICH) 149
National Lands Advisory Committee 68
National Marine Forum 93
National Parks Commission 51
National Parks System Act 51
National Protected Areas Policy and System Plan Project 57
Natural Heritage 93
Naturalight Productions 119
Natural Resources Defence Council (NRDC) 57, 67
Natural Resources Management Project (NARMAP) 18, 84, 85, 164, 229, 256
Natural Resource Week 93
Nature Conservancy, The (TNC) 30, 110, 128, 132, 134, 257, 146, 198, 253, 255, 257
NBC 119
NEAC 61, 67, 68, 75, 80, 229
Neal, June 13, 83, 142, 148
Neal, R. H. 25
Nealey, Rebecca 136
New York Zoological Society 57, 58, 125, 159, 255, 258
Newsletter, BAS 18, 66, 67, 83, 107, 119, 179
NICH 149
Nicolait, Lou 4, 209, 226, 227
Nicolait, Robert 7, 227
Noble, Delia 38
Nolan, Mark 18, 256
Northern Drowned Caye 75, 79
Northern Lagoon 134, 178, 182
Northern Lagoon Manatee 134, 212
NRDC 57, 67

O

Oak Foundation 30, 117, 220, 243, 253, 257
observation tower in Crooked Tree 154
observation tower in Half Moon Caye 125
office 11, 35-38
office staff 14
offshore drilling for oil 78
oil exploitation 77-78, 107, 229
oil refinery in Monkey River 77, 80
Old Belize 41
Old Woman Creek 236
Olewin, Ben 232
Olive-backed Euphonia 142
Orang-utan Press 109, 111
orchid 47, 104
Ostro, Linde E. T. 185, 186, 189

P

PACT ix, 30, 34, 41, 48, 79, 117, 118, 128, 137, 143, 220, 229, 243, 255, 257
Palacio, Andy 94
Palotti High School Nature Protectors 41
Panthera 118
paraquat 57
Paredes, Alejandro 178
Paredes, Humberto 18, 19
Parker, Jim 2, 5, 125
Parrot, Yellow-headed 53
Parrot Heliconia (*Heliconia psittacorum*) 47
parrots 52, 53, 54
Partners 227-266
Pau, Marcelo 31, 88, 190, 192
Pau, Renny 190
Pavek, Richard 50
Paxz Tours 148

Peace Corps Volunteer 5, 23, 83, 108, 110, 135, 136, 140, 146, 152, 155, 160, 164, 194, 228, 240, 246
Pearl, Mary 224
Pelican Beach Resort 85, 88
Pendergast, David 108, 160
Pendergast, Esther 2, 3
People's United Party 65
Peppermill Casinos, Inc 119
Perdomo, Carlos 237
Perez, Arlenie 263
Perez, Pedro 13
Performance Bond 68
Periott, Tremett 48
Pesticide Control Act 57
Petroleum Act 78
Phalaenopsis amabilis 45
Philip, H.R.H. Prince 160-1, 222
pier at Half Moon Caye 128, 257
Pine Lumber 35, 231
Pixabaj, Pedro 189
Placencia 75
Plant Inventory in CBWS 192-3
Police 41, 147, 168
Policy and Operations Manual 34
Pook's Hill Lodge 147
Pop, Bernarda 165
Pop, Celestina 165
Pop, Emiliano 189
Pop, Estevan 165
Pop, Ignacio 31, 222
Pop, Rosita 24
Population 107
Port Honduras Marine Reserve 174, 179, 214
Posters 115-116
Pott, Amador 199, 200
Pott, Roberto 31, 56, 130, 171, 180, 195, 198
Prance, G.T. 192
Premier of Belize 3
Premier's Lodge 49
Price, George 1, 2, 3
Programme for Belize 12, 49, 54, 59, 63, 112, 162, 195, 227, 228, 229
Protected Areas 7, 14, 26, 107, 123-170
Protected Areas Conservation Trust (PACT) ix, 30, 34, 41, 48, 79, 117, 118, 128, 137, 143, 220, 229, 243, 255, 257, 255
Protected Areas Manager 14
Publicity Coordinator 14
Punta Allen, Quintana Roo, Mexico 249

R

Rabinowitz, Dr. Alan 159, 184, 186
Radio Programmes on Radio Belize 82
Radisson Fort George Hotel 18, 19, 20, 62, 64, 93
Rainforest Exhibit 93
Raleigh International 141, 147, 240
Ramsar Convention 72, 99, 153, 170, 258
Rare Animal Relief Effort (RARE) 84, 99, 142, 231
RARE 84, 99, 142, 231
Rath, Therese Bowman 11, 27, 28, 33, 36, 39, 160, 162, 208, 211, 226, 256, 261
Rath, Tony 115
Red Bank 54
REDES-Central America 237
Red-footed Booby 8, 49, 124-5, 177, 204, 251,
Redington, Kerrien 110
Reed, M.L. 184
Reef Preservation Committee viii, 55, 80, 109, 112, 114, 115, 125, 205, 207, 241
Regent Insurance Company 12
Regent Street Office 35
Reneau, Annette 30, 32
renewable energy 107
reptiles 191
Research 7, 14, 32, 82, 171-204, 236
Retreats 26-28
Rhaburn, Uriah 33, 88
Rich, Bruce 57
Rio Bravo Conservation and Management Area 29, 59, 174, 228
Riverol, Bardi 27, 28, 237
Roadside Hawk 142
Robateau, J. J. 125
Robinson, Julianne 14, 112, 127, 195, 197
Rosado, Gil 1, 2, 4, 7, 82
Rosado, Harvey 31
Rosenzweig, David 91
Rossington, Anna 20, 112, 119, 211
Roupala glaberrima (*Proteaceae*) 192
Rubido, Ambassador Manuel 42
rushes (*Eleocharis cellosa* and *E. interstincta*) 182

S

Sabido, Wilber 100
Sacred Heart College 226
Salas, Osmany 13, 18, 20, 27, 28, 40, 84, 87, 158
Samuels, Verna 239
Santos, Carlos 19, 22, 27, 28, 30, 40,
Saqui, Claudia 165
Saqui, Ernesto 87, 184, 160, 188, 189, 209, 213, 226, 237, 265
Saqui , Hermelindo 189
Saqui, Pio 88, 183
Sarstoon-Temash National Park 59
Save Our National Heritage 78
Save our Rivers Campaign 213
sawgrass (*Cladium jamaicense*) 182
Scarlet Macaws 54, 190
Schipp, William A. 192
Schmidt, Dr. Peter 2, 4
Schmidt, Scott 23, 30, 32, 108, 194, 240
Schutte, Carol 184
Scott, Jr., Patrick 223
Scott, Sr., Patrick 19, 20
Scouts 41, 82, 241
Searle, Linda 20, 22, 31, 116, 163
sea grass 115, 198, 199
sea turtles 55, 80, 93, 112, 198, 199, 201, 204
Shapley, Rebecca 184, 196, 202-4
sharks 55, 116, 197,
Sharp, Francesca 221
Shaw, Jean 40, 165
SHBHNP 89, 105, 106, 140-145, 194, 235, 239, 242, 247, 255
Shipstern Nature Reserve 228
Sho, Celso 194

Sho, Eligorio 189
Sho, Emiliano 223, 265
Sho, Timothy 192
shoal grass (*Halodule wrightii*) 199
Shuey, John 184
Shutter Speed 42-7
Sibun Forest Reserve 235, 239
Sibun Watershed Association 210, 220, 226, 229
Silver, Scott C. 185, 186, 189
Simmons, Wilmot 18
Singh, Michael 18
Skeen, Jasmine 221
slideshows 117
SMART 41
Smith, Amy E. 110, 146
Smith, Greg 55, 207, 226
Smithsonian Institute 57-8
Snake Cayes 179
Snakes of Belize 109
Social Flycatcher 142
Society Hall Nature Reserve (TMNR) 10, 110, 146-147, 239, 257
Society for the Promotion of Education and Research (SPEAR) 63, 68
Somerville, Michael 14, 40, 86, 136
Sosa, Lucilo 25
Sosa, Nigeli 142
South Beach Development 76
South Stann Creek 158, 189
Spanish Bay Development 76
spawning aggregation 195, 198, 202, 204, 214, 230, 257, 266
SPEAR 63, 68
Special Constable training 87
Special Expeditions 115
Sprunt IV, Alexander 150
Staine, Albert 1, 2, 4
Staines, Norman 1, 2, 5, 82, 83
Stake Bank 75, 79
Stakeholder Communities 243-252
Stella Maris School 96, 97
St. Herman's Blue Hole National Park (SHBHNP) 89, 105, 106, 140-145, 194, 235, 239, 242, 247, 255
Caribbean whiptail stingray (*Himantura schmardae*) 202
St. John's College First Aid Team 41
Strategic Planning 26-32
Strengths, Weaknesses, Opportunities, and Threats (SWOT) 29-31
Striped Basilisk (*Basiliscus vittatus*) 146
Students Encouraging Environmental Development (SEED) 241
Students Understanding Nature (SUN) 241
Submontane Elfin Shrubland 190
Summer Camps 15, 104-5
Summer Programme in Tropical Biology 82
Summit Foundation 253, 258
Sun Jam 56
Swallow Caye Wildlife Sanctuary 75, 174, 209, 212
Swaso, Major G. H. 56
Swim for the Reef 109, 241
SWOT 29-31
Szabo, Paul 2, 5

T

Talisia macrophylla (*Sapindaceae*) 192
Tapir Mountain Nature Reserve (TMNR) 10, 25, 83, 84, 110, 123, 146-148, 170, 180, 194, 239, 240, 242, 253, 257, 266
Target Earth 235
Tarpon (*Tarpon atlanticus*) 183
Taylor, Daniel 184
Teacher's Day 90
Teacher Training 90-92
Teck, Sarita 245
Tern, Sooty 179
Tern, Roseate 179
Terra Foundation 111
Teul, Mario 22, 230, 263
Teul, Pulcheria 215, 226
TIDE 198, 214, 227, 255
Tiger Fern Trail 46, 168
Tilapia 69, 70, 101, 183, 204
Tillett, Ambrose 18
Tillett, Arlita 33
Tillett, Donald 70, 88, 99, 228
Tillett, Lascelle 18, 19, 20, 22, 43, 106, 263
Tillett, Nancy 19, 119
Tillett, Sarita 23, 43
Tillett, Steve 33, 103,152, 180, 182, 183, 222, 224, 243, 265
TMNR 10, 25, 83, 84, 110, 123, 146-148, 170, 180, 194, 239, 240, 242, 253, 257, 266
TNC 30, 110, 128, 132, 134, 257, 146, 198, 253, 255, 257
Toledo Institute for Development and Environment (TIDE) 198, 214, 227, 255
Tour Guide Training 85
Training 87-88
Trekforce Expeditions 154, 168, 242
Trinidad Farm 27
Trogan, Black-headed 142
Tropical Flora and Fauna Preservation Society 111
Tropical Research Centre 82
Tropical Storm Alma 20
Tropical Storm Arthur 20
turtle grass (*Thalassia testudinum*) 199
Turtle Week 93
Tush, Federico 33, 265
Tush, Santos 190
Tzul, David 146

U

UK Center for Ecology and Hydrology 181
UNESCO 18, 126, 128, 132
UNDP 54, 85, 195, 217, 246, 258
United Democratic Party 65
United Nations Development Programme (UNDP) 54, 85, 195, 217, 246, 258
University College of Belize 39, 88
University of Belize 41, 83, 103, 132, 181, 183, 191, 195, 198, 220, 237
University of Belize Environmental Club 41
U.S. Agency for International Development (USAID) 18, 33, 57, 85, 256, 257
USAID 18, 33, 57, 85, 256, 257
U.S. Embassy in Belize 41
U. S. Fish and Wildlife 55, 111,114, 137

Representative Stanley Usher 162
Ushia 119

V

Vaca Dam 68
van Opstal, Marcel 18
Vasquez, Ernesto 16
Vasquez, Marvin 181
Vasquez, Virginia 12, 13, 164
Vega, Hon. Gaspar 98, 226
Venture Scouts 241
Verde, Elvia 2
Verde, Jorge 2, 4
Verner, Jared 177
Vernon, Leela 41, 97, 98, 118, 218
Victoria Peak Natural Monument 123, 169, 190-1, 241,242
Village Council 18, 251
Vireo 111
Volunteer of the Year 223

W

Wade-Moore, Diane 76
Wagner, Gilda 2
Waight, Christopher 215, 219
Waight, Daniel 216, 219
Waight, Gerard 212, 219
Waight, Gloria 216
Waight, Hyacinth 2
Waight, James A. 1, 51, 159, 205-221, 237, 261
Waight, John 216, 219
Waight, John William 211
Waight, Lydia x, 1, 51, 81, 83, 107, 112, 122, 140, 150, 171, 237
Waight, Maria Teresa 213
Walkathon 40
Walker, Paul 152, 190-1
Walk for a Green Belize 41, 98
Warden of the Year 222
waste management 65
Water Quality Monitoring Programme 182
Water Resources 61, 65, 76, 95, 107, 125, 182
Waters, Russell 3, 82
WCS 30, 185, 186-7, 188, 195, 202-4, 253, 258-9
Weber, S.J., Fr. Leo 1, 2, 4
website 119-121
Wechsel, Mark 186
Weeden Foundation 258
Westby, Joy 97
Western Lagoon 71-2, 151
Westrom, Dr. Wendy 188
Wetlands Day, World 99-101, 121, 122, 257
Wetlands Summer Camp 92
Weyer, Diane 8
Weyer, Dora 1, 2, 5, 6, 7, 39, 82, 108, 109, 125, 134, 150, 159, 205, 206, 207, 217, 225, 226
whale sharks 197
Whaling Commission 55
White-throated Sparrow (*Zonotrichia alicollis*) 234
WIDECAST 55
Wider Caribbean Sea Turtle Conservation Network 55
Wildlife Conservation Society 30, 185, 186-7, 188, 195, 202-4, 253, 258-9
Wildlife Protection 52-56
Wildlife Protection Act 51
Wild Wings Foundation 13, 150, 258
Williams, Carol 30, 33
Williams, Dennis 180
Williams, Diana 128
Williams-Thompson, Tanya 61, 63, 69, 79
Windevoxhel, Nestor 18
wish-willies (*Ctenosaura similis*) 177
Wood, D. Scott 108
Woods, S.J., Fr. Charles 1, 2, 4, 18
Woods, Sylvana 18
Woods, Valerie 40, 210
World Bird Count 175
World Bird Festival 176
World Conservation Union (IUCN) 27, 30, 237
World Gap Challenge 241, 242
World Heritage Site 113, 117, 126, 132, 170
World Ocean Day 78
World Wetlands Day Celebrations 99-101, 121, 122, 257
World Wildlife Fund (WWF) 41 55, 109, 222, 259
WWF 41 55, 109, 222, 259

X

xaté palm (*Chamaedorea ernesti augustii*) 149, 194, 204
xatéros 149
Xavier Primary School 84

Y

YMCA Camp 105
Young, Carlton 20, 48, 100, 263
Young, Fallett 237
Young, Governor General His Excellency Sir Dr. Colville 48, 100, 219
Young, Shane 33
Young, W. Ford 1, 2, 4, 5-6, 7, 17, 82, 83, 108, 124, 140, 146, 171-2, 178, 206, 207, 217, 225, 226, 232
Youth Education Committee 2, 49, 81-82
Youth Enhancement Service 104
Youth Environmental Action Group (YEAG) 241
YWCA Camp 104

Z

Zoological Society of Milwaukee 230

www.ingramcontent.com/pod-product-compliance
Lightning Source LLC
Chambersburg PA
CBHW041320210926
51847CB00002B/9
9789768142290